Cryptoasset Inheritance Planning
A Simple Guide for Owners

Pamela Morgan, Esq

Table of Contents

Praise for Cryptoasset Inheritance Planning

> Pamela provides an easily digestible guide to protecting your cryptoassets from being lost upon your passing. I've spent many hours working on my own plans but was still surprised by several potential pitfalls that were covered. Anyone who owns cryptoassets should read this book, even if they're highly technical or expect to live for many years.
>
> — Jameson Lopp, Software Engineer at Casa

> This excellent guide demystifies a challenging and complex issue that has not received sufficient attention to date: the issue of estate planning for those who own cryptoassets. The guide is accessible, practical, thorough and highly readable. More broadly, it is a useful guide for anyone seeking a plain language discussion of security issues surrounding ownership of cryptoassets, particularly those in the legal profession.
>
> — Ana Badour, Partner and Fintech Lead at McCarthy Tétrault LLP

> Cryptoasset Inheritance Planning: A Simple Guide for Owners is a must have for folks who want their loved ones to be the beneficiaries of their crypto holdings. This book provides simple, easy to use instructions that will give you a concrete plan to maximize the chance that your Bitcoin, Ether and other coins get to the people you want to receive them. Cryptoasset Inheritance Planning is simply the best resource for anyone from the average Joe to the crypto expert.
>
> — Daniel Serlin, Managing Partner at Serlin Trivax & Associates PC

> Valuable, practical and timely, I expect that this is a book that I will read more than once and recommend many times. You don't have to be a lawyer to read this book, but every lawyer that works in estate and wealth planning should read it.
>
> — Amber D. Scott, compliance nerd / bitcoin enthusiast & Founder of Outlier Solutions

Colophon

Pamela Morgan, Esq.

https://empoweredlaw.com/

@pamelawjd

Cover Design

Kathrine Smith: http://kathrinevsmith.com/

Review and Editing

Jessica Levesque: https://charlestreellc.com/

Walter the Unsalter: http://cleanwatermn.org/road-salt-less-is-more/

Copyediting

Reuben Thomas

rrt@sc3d.org

@sc3d

First print: **May 1, 2018**

Errata Submissions: errata@merklebloom.com

Licensing Requests: licensing@merklebloom.com

General: info@merklebloom.com

ISBN: 978-1-947910-11-9

Disclaimers:

This book is edited commentary and opinion. Much of the content is based upon personal experience and anecdotal evidence. It is meant to promote thoughtful consideration of ideas, spur philosophical debate, and inspire further independent research. It is not investment advice; don't use it to make investment-related decisions. Despite being written by a lawyer, It's not legal advice; consult a lawyer in your jurisdiction with legal questions. It may contain errors and omissions, despite our best efforts. Pamela Morgan, Merkle Bloom LLC, editors, copy editors, reviewers, and designers assume no responsibility for errors or omissions. Things change quickly in the cryptoasset industry; use this book as one reference, not your only reference.

References to trademarked or copyrighted works are for criticism and commentary only. Any trademarked terms are the property of their respective owners. References to individuals, companies, products, and services are included for illustrative purposes only and should not be considered endorsements.

Licensing:

Foreword

By Andreas M. Antonopoulos

> If half of what this guy says is true, this technology could change the world.
>
> — Pamela Morgan, Disrupt 2013

In November 2013, Pamela Morgan was in Athens, Greece, speaking at the "Disrupt 2013" conference. She was teaching a workshop on entrepreneurship and the opportunities it offers for people with disabilities. By coincidence, on the second day of the conference, I presented a keynote talk about bitcoin and its revolutionary potential. Pamela remembers her reaction to the 20 minute talk: "If half of what this guy says is true, this technology could change the world." Like many of us, Pamela had just had her crypto-epiphany.

People get involved in bitcoin and crypto-currencies for different reasons, but I often hear a common story. After hearing the word "bitcoin" a few times, many people experience a moment of clarity in which all the implications and possibilities suddenly become apparent. *Epiphany* is a Greek word describing this moment of insight and sudden understanding. If you're reading this book, you likely had that exact experience at some point in the last 9 years.

After an epiphany like this, it's hard to carry on as usual. The ideas and possibilities keep swirling in your head and can be very distracting. My moment of epiphany came in 2012, a bit more than a year and a half earlier. I had read about bitcoin before, but had dismissed it as "nerd money for gambling". But this time, the article I read had a link to the white paper by Satoshi Nakamoto. This time, I couldn't dismiss it. I became quite obsessed.

A year after first meeting Pamela, she invited me to join Third Key Solutions, a company she formed to explore key management, multi-signature and governance solutions. She had already abandoned her full-time job and had spent a year learning everything she could about crypto-currencies.

Pamela is an educator, attorney and entrepreneur. She's uncompromising in her ethics and focused on making the world a better place. Now, that sounds like a cliche, but once you meet and work with Pamela you quickly realize that it guides her every decision - she lives what she believes. She's taught me a lot by example.

Pamela studied Business and Computer Programming before pursuing a Juris Doctor (JD) degree. She never really wanted to practice law in a traditional sense. Rather, a law degree was a way to empower herself and others through her work. Most attorneys are unfortunately so focused on law that they forget about justice, but Pamela does the opposite: she views law only in the context of justice. If it doesn't work, it's time to move on and build something better.

This attitude drives her to see the world as interconnected "systems", that either serve their declared purpose... or don't. If they don't they must be re-engineered or discarded. Unlike most attorneys she's not afraid of technology, but rather is extremely adept at learning and using it to improve the systems that govern our lives. She's truly part of a new generation of cross discipline experts who effortlessly bridges technology and law. Exactly the kind of skill set needed in the very important topic explored in this book: cryptoasset inheritance planning.

At Third Key Solutions, we work together to build processes and tools that empowered our clients to use crypto-currencies with maximum independence. We do not hold custody of funds or direct our clients to trust third parties. Instead, we teach them how to take control and responsibility over their own assets and their own security.

Inevitably though, the highly personal and customized services we offer through Third Key Solutions do not scale. Demand significantly outpaces supply. To scale, Pamela decided to take the experience and knowledge accumulated over four years and distill it into a "simple guide" that anyone could apply. That's how this book came to be.

In "Cryptoasset Inheritance Planning - A Simple Guide For Owners", Pamela takes all the technical, legal, procedural and risk management knowledge built through her work with dozens of clients and makes it easy for everyone to apply. This is not a book about theory, complex math, or law. It's a practical, pragmatic, and applicable step-by-step guide. This book has a single focus: making it possible for a non-technical person to convey their cryptoassets to non-technical heirs.

Pamela's first book is balanced, detailed and practical. I'm about to finish my fourth book, but I remember clearly how difficult it was to produce the first one. It's hard work, emotionally exhausting, and intellectually taxing but ultimately rewarding.

You should really think about "doing" this book, rather than reading it. It's almost like a workshop on paper. If you follow the practical steps in this book, you will end up with a plan with a high likelihood of success. Your assets will be more secure, you will learn a lot and your heirs will have a great chance of benefiting from your generosity and prescient investment.

Get it done!

Andreas M. Antonopoulos

Andreas is a public speaker, CTO at Third Key Solutions , and author of three widely acclaimed books about open blockchains and crypto-currencies: "Mastering Bitcoin (2nd Edition)", and "The Internet of Money" Vol 1 & 2. He's currently working on his fourth book: "Mastering Ethereum."

Introduction

Who Is This Book For?

If you own any cryptocurrency or cryptoasset tokens, this book is for you. If you use an exchange to buy and sell cryptoassets, this book is for you. If you've ever wondered or asked the question, "What will happen to my bitcoin, ether , or other cryptoassets when I die?" this book is for you. If you want someone, anyone, to inherit your cryptoassets when you die, this book is for you.

Perhaps you, like most people, have never thought about what will happen to your bitcoin, ether, and other cryptoassets when you're no longer around. It's not the most fun thing to contemplate and it's hard to know where to start. Even if you do understand where to start and what needs to be done, actually doing the work to make it happen is often a daunting prospect. It's one of those tasks, without a hard deadline, that gets pushed to the end of your to-do list, repeatedly.

The significant challenge is to maintain security and privacy while at the same time ensuring that all the keys required to access your cryptoassets will be available to your heirs at, and only at, the appropriate time. This book will guide you through the process of building a comprehensive access plan for your cryptoassets. There are several simple, easy things you can (and must) do to make it more likely that your loved ones will be able to inherit these assets when the time comes. The first and easiest step was buying this book. The next step, follow-through, requires thought and diligence. This book is designed to help you create a plan that will work for you and take into consideration your unique situation. There is no cookie-cutter plan; no fill-in-the-blank or downloadable form can do this for you. But don't worry, I've helped many people through the process of building customized plans, and I've put those ideas and experiences into this book. Through this book, I will help you too.

This book isn't written specifically for lawyers, security experts, or cryptographers, though they may benefit from the material. The entire book, with the exception of the Making it Legal, is applicable to any cryptoasset owner in any jurisdiction. The Making it Legal section cites some USA law but the principles are broadly applicable around the world.

Will this book teach me about specific cryptoasset laws in my jurisdiction? No. A book like that is called a "legal treatise;" they're heady and dense, even for lawyers. Instead, this book focuses on practical information you need know, like what happens to your assets if you don't have a will and why you shouldn't put your cryptographic keys in your will. You'll learn about high-level legal concepts that might affect your assets, how to find out more information about the laws in your jurisdiction, and how to keep legal costs down.

The unique challenges with cryptoasset inheritance planning are not primarily legal—they're primarily technical. With this book, you'll learn how to create a cryptoasset access plan for your heirs. Your access plan will aim to answer the question, "From a practical perspective, how will my loved ones access my cryptoassets when I'm not around to help them?"

Goals

This book is best read with four goals in mind:

- Allow your heirs to take possession of your cryptoassets when the time comes, but not before.
- Minimize the risk and opportunity for anyone to steal cryptoassets before they're delivered to your loved ones.
- Provide the opportunity for your loved ones to hold the assets securely, if they choose to do so.
- Prevent disputes amongst your heirs and avoid legal problems whenever possible.

By the time you've worked through the first few chapters of

this book, you will have completed a Get It Done plan. If you don't at least complete that part of the planning, I've failed. My job, the intent and purpose of this book, is to motivate you, to inspire you, to help you actually create your plan. These things don't happen by osmosis; just having the book near you won't be enough to protect your loved ones. If you don't take action and develop a written plan, your heirs probably won't inherit your cryptocurrencies and tokens. The Get It Done plan is the easiest, fastest way to get started.

My hope is that you'll go beyond the Get It Done plan. Ideally, by the time you've finished this book, you'll have a comprehensive access plan (the technical side) and you'll be working to create a distribution plan (the legal side) for your cryptoassets. You might even decide to expand your plan to include other digital and traditional assets. Bottom line, by reading this book, you'll understand how to improve and maintain your access plans, how to choose people to help your heirs with the process, and how to inform your loved ones about your assets in a way that they can understand. You'll learn about legal issues, like wills and trusts, how to find and hire a knowledgeable lawyer, how to keep costs down, and how to self-educate. Additionally, you'll learn how and why you need to keep your plan up to date so that your loved ones aren't left with stale, outdated information.

Planning is For You Too

Inheritance planning isn't just for your loved ones; you'll benefit as well. Creating a comprehensive plan will require you to back up everything and you'll know where each backup is located. You'll think about security, usability, resiliency, and efficiency. And you'll have a contingency plan that identifies what to do if something happens to your access devices like your laptop or phone. You'll have peace of mind for your loved ones and for yourself.

Perfect Solutions are an Illusion

There is no such thing as perfect security, just as there is no such thing as zero risk. The processes in this book all work to balance security and accessibility. You'll spend some time thinking about the risks you actually face and find ways to mitigate them. There is no such thing as a perfect process that is guaranteed to work when the time comes. That's true for all inheritance planning, not just for cryptoassets. For example, even with a thorough plan your heirs still might not be able to inherit your assets because of a natural disaster or something else outside of your control. But for the vast majority of people, in the vast majority of situations, a well-designed, well-executed, current, tested access plan will provide enough information to allow your loved ones to inherit your cryptoassets. And that is our primary goal.

This book is written for everyday people, who want financially and technologically reasonable solutions to their cryptoasset inheritance problems. It's not about building the most secure, inaccessible Fort Knox storage solution (that your loved ones would never be able to use). You won't need an airgapped laptop with the microphone and cameras physically removed. You won't need to become a security expert. But you will learn more about security, backups, and you might even learn something about yourself and your values in the process.

There is no right way to do this planning; there is only the right way for you and your situation. In this book you'll learn what I've learned from helping my clients and my years of trial and error experience. If you follow the steps, you should end up with a plan that works well for you. Using this book will help you to avoid planning errors and traps along the way. I'll help you avoid the most common mistakes and you'll think about inheritance planning in a whole new way.

How to Read This Book?

There are five major sections in this book. The first outlines the six most common mistaken beliefs that stop us from planning. Then you'll move right into Get It Done, and as the name implies it's all about putting a plan together right now. Today! All you need are pen and paper and an hour or two to read the step-by-step instructions and get a basic access plan in place. In a few hours, you'll have accomplished the first step in building your comprehensive inheritance plan, and your heirs will be significantly more likely to inherit your cryptoassets.

The third section is all about Making It Better, 'it' being your first plan. In this section you'll learn how to make your plan more secure, resilient, usable, and efficient; we'll talk about common misconceptions and pitfalls and how to avoid them. The fourth section is about Making It Legal. We'll talk about why you might want to add a legal component to your plan and how to do it, inexpensively and easily. In section five, we'll discuss Keeping it Fresh: how to keep your plans updated.

The book is designed to be a reference guide too. Look to the table of contents to find specific sections when you need them. You'll also find blank templates in the appendices and resources sprinkled throughout the text.

Disclaimers and Clarifications

Nothing in this book should be considered legal advice. The contents of this book are intended to be a starting point for cryptocurrency and cryptoasset holders to understand inheritance. For specific questions regarding your legal inheritance plan, speak with an attorney in your jurisdiction who understands your particular needs and goals.

Mentions of specific software, hardware, and services are not endorsements. They are used as examples only, and they should not be considered as more or less secure or reliable than any other software, hardware, or services. Do your own research.

This book is written for laypeople, not lawyers. The words used here are meant to convey their everyday meaning and should not be considered legal terms of art unless otherwise indicated.

This book is written for laypeople, not security experts. It's written for people with limited to basic experience in operational and information security. If you're an expert, feel free to adjust your plan to reflect your expertise while remembering that your heirs probably don't have your level of skill or knowledge.

This is not a how-to-do-security book; it is a book about providing access to your loved ones when you're not around to help them, and that necessarily involves processes that balance security and accessibility.

Following the advice in this book will not guarantee that your heirs will be able to inherit your cryptocurrencies and cryptoassets. But not creating an access plan for your loved ones almost certainly means they won't be able to inherit these assets. You will need to decide how to apply the advice in this book based upon your specific situation.

Independent Recovery Strategy

With high-quality plans, your loved ones will not need access to your devices to access your assets. You will back up the cryptographic keys, seeds, and access codes themselves on paper or standalone electronic media, so that the assets can be recovered without the devices you use to access your assets, like your smartphone and laptop. There are many benefits to such a strategy, including maintaining your privacy. I don't want my loved ones poring over my browsing history or reading all of my stored messages. You probably don't either. But if you do, you may wish to also develop a plan for recovery of the data on your devices, such as smartphones and laptops, and your online accounts, like Twitter and Facebook. That strategy needs to go beyond leaving passwords for your heirs and is beyond the scope of this book. For cryptoasset recovery, device-independent recovery is a much better choice.

Six Mistaken Beliefs

If you're reading this book, it means you know you need to plan. If you're ready to make that plan now, you can skip to the next section. If you're not sure why you need to plan, read on!

In February of 2018, I posted a Twitter poll, asking about inheritance planning for cryptoassets. More than 5,000 people voted, and, to my dismay, 55% of people said they aren't planning for these assets to pass to their loved ones.

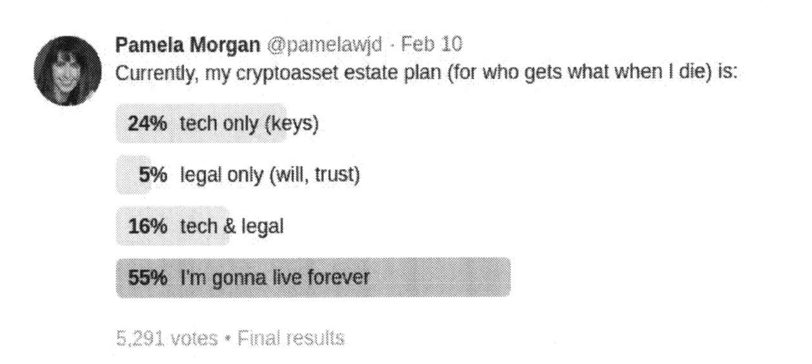

Figure 1. Twitter Poll: We plan to live forever

While people have lots of reasons, or excuses, for why they don't plan, I believe the real reason is denial and fear. We don't like to face our mortality. For the young, death often doesn't seem real. No matter your age, health, or life circumstances, it's not pleasant to focus on dying. And planning for our own death forces us to think about it for more than a minute. We seem to believe that if we don't think about dying it won't happen.

In addition to facing our fear of death, the topic of inheritance planning for cryptocurrencies and tokens is fraught with misconceptions. Some of these misconceptions are due to the technical nature of cryptoasset inheritance planning. Some are due to the legal complexity of inheritance planning in general.

Over the past several years, I have spent much of my time writing and teaching about inheritance planning for cryptoassets. At first I was surprised at the level of resistance I faced. Even though this is something everyone with cryptoassets should do, it was surprisingly difficult to persuade people to do it! Every article, talk and even tweet was immediately followed by a flood of objections. By studying these objections, I have found that most of them are related to the following 6 misconceptions.

1. Mistaken Belief: I have to hire a lawyer.
2. Mistaken Belief: I have to trust a third party.
3. Mistaken Belief: Planning will make my assets easy to steal.
4. Mistaken Belief: The value of my cryptoassets is too low to plan.
5. Mistaken Belief: My heirs will figure it out.
6. Mistaken Belief: This can all be done with a simple smart contract.

Mistaken Belief #1: I have to hire a lawyer

There are two sides to planning for cryptoassets or digital assets we control: the technical access plan (access to keys, seeds, and required access controls like passphrases) and the legal distribution plan (local laws, jurisdiction, who gets what). Without access to keys, seeds, and other access controls, legal distribution plans are pointless, because there is no practical way to carry them out.

Most lawyers have no idea what a private key is, and they do not understand key security. They cannot help you design a plan for securing your cryptocurrency assets during your lifetime while providing access to them only upon death. But that doesn't mean you don't want a lawyer to help with the legal side of things, like writing your will and making sure your wishes will actually be carried out. My point is that you don't need to wait to hire a lawyer to design your key access strategy. It's best for most people to start with an access plan and add a legal component later, and that's what we'll do in this book.

Mistaken Belief #2: I have to trust a third party

The point of owner-controlled assets is to not give others power over them. The point is to maintain control over them yourself. The plans in this book will *not* require you to hand over your keys to any third party. There are creative ways to design key access plans that will enable access by others in the future, without giving them immediate access. While these types of solutions are often uniquely designed for the person and their specific situation, an example of a simple option you can use today is to add a passphrase to your Trezor or Ledger hardware wallet device. We'll examine this concept in detail later in the book but for now it's enough to know that it's possible to distribute control in a way that doesn't give any one person complete power.

Sometimes we forget that a third party may be a software agent—like that new estate planning ICO (Initial Coin Offering) you may have heard about recently. Be wary of any solution that requires you to put all your trust in a single third party, especially if that third party is not a known individual who is legally responsible to your heirs, as in the case of a software agent written by a team of developers.

Mistaken Belief #3: Planning will make my assets easy to steal

A good plan does not, and should not, require you to list all of your assets and access passwords and store them in one place. The point of good inheritance planning is to make the assets accessible to others only when they have the right to them and not before. And even if you did write down all of your credentials on a sheet of paper, the truth is you're much more likely to lose your assets by failing to back up your wallet, leaving money on an exchange, losing your backup or paper wallet, or forgetting your password/passphrases. That said, it would not be smart to store a single document with all of your passwords, passphrases, access points, holdings and amounts online or on your internet-connected device. But you already knew that. You probably already know to handwrite these things down on paper, store them in a secure, locked location, and to also use a password manager. We'll touch on secure storage a bit later in the book. Remember though, your inheritance planning isn't dependent upon you performing a security makeover. Save that for another day and keep reading.

Mistaken Belief #4: The value of my cryptoassets is too low to plan

While this might be true for you today, the value of these assets can change very quickly. For example, in April 2017, the price of bitcoin began a steep rise from about $1,200, reaching a peak of $20,000 per coin in December 2017. In January 2018, the price was hovering near $12,000, and in mid-March 2018 the price is hovering around $9,000. Similarly, in April 2017, ether was trading at $50, but by the end of December it was near $850. In January 2018, it was hovering near $1,200 and in mid-March it's around $800. If we can learn one thing from this volatility, it's that our planning shouldn't be dependent on fiat exchange price. Small holdings might quickly become significant. You don't know what the value will be in the future, but you're holding on to what you have now for a reason. That's reason enough to plan.

Mistaken Belief #5: My heirs will figure it out

You know that's not true. We all know that glazed-over look our loved ones get when we start talking cryptoassets. There are really only two scenarios where they are able to figure it out on their own: (1) if your security is so terrible that anyone can figure it out, in which case your assets will probably be gone long before you, or (2) you're in that rare unicorn position where your loved ones are as nerdy as you are. For most people, if you don't provide an access plan, your assets will perish with you. Your heirs will probably throw away, factory reset, or donate the devices that have your keys on them. They'll ignore or throw away private keys because they won't know what they are or how to use them, and won't understand their value. And if they ever figure out what they've missed out on, it will be too late to do anything about it. We'll talk more about educating your heirs later in the book.

Mistaken Belief #6: All I need is a smart contract

This can all be done with a simple smart contract, deadman's switch, electronic will, digital oracle, and a sprinkling of artificial intelligence. This isn't actually a misconception; it's the marketing plan of a dozen startups and it has successfully distracted many from the realities of pragmatic cryptoasset inheritance planning today. These ideas have become so pervasive and engender such risk that they require a thorough debunking and a strong dose of reality in a dedicated chapter. To avoid further distraction, it's been moved to an appendix of the book Smart Contracts For

Inheritance Planning.

A deadman's switch can be a useful mechanism to notify your heirs about the existence of your cryptoassets, but not convey any confidential information. This concept is covered in When a Dead Man's Switch Might Be Useful.

Get It Done: Making Your First Plan

Over the next hour or two you will write a basic access plan for your heirs. It will be quick, easy, and probably incomplete. But it will significantly improve the chances of your heirs actually inheriting some of your cryptoassets. Your access plan will be written as a letter to your heirs, and it will cover just the basics: who can help them, where to find the assets, which devices they should preserve, and where to find more information. It will also caution them not to trust any one person.

Remember the four goals (in order):

- Allow your heirs to take possession of your cryptoassets when the time comes, but not before.
- Minimize the risk and opportunity for anyone to steal cryptoassets before they're delivered to your loved ones.
- Provide the opportunity for your loved ones to hold the assets securely, if they choose to do so.
- Prevent disputes amongst your heirs and avoid legal problems whenever possible.

This letter is meant to be an interim measure; a stop-gap. It is intended to provide your loved ones with something, not everything. In the next section of the book, Making it Better, we'll create a more detailed and thorough plan, but for now let's get this done.

Supplies Checklist

In order to complete your plan, you'll need the following supplies:

- ☐ Blank paper: 4–6 pages, at least ½ sheet size
- ☐ Ink Pen
- ☐ Envelopes: 2–3
- ☐ Your phone or address book
- ☐ Your laptop or desktop computer

Optional: new USB device

Optional: printer

Optional: wallet backup cards

Optional: use the worksheets in the back of the book or join our mailing list and receive PDF copies to print yourself

Why Those Supplies?

Old-fashioned paper and pen? Can't I just use my phone or computer? For this exercise, we assume that your internet-connected devices have a virus, or will have one someday. While you won't be storing any keys or access credentials in the document, why leave a digital trail when you don't have to? It's good to get in the habit of using pen and paper, particularly for cryptoasset storage.

You'll want to use pen rather than pencil because ink is more stable. You don't want an heir who finds the plan to be able to erase some of the assets or devices. Even if you believe your heirs wouldn't do something like that, if you erase something, other heirs might think the person who found the inventory erased something. For these reasons, it's best to use permanent ink.

You'll need multiple pieces of paper because at the end of this exercise you'll copy your completed letter onto two or three sheets of paper and store them securely. You might also need to back up a seed, password, or passphrase on paper.

You'll need a new USB device only if you use software that utilize keyfiles as backups — for example, Mist or a Monero full-node wallet. Be aware that most of these also require a password in order to decrypt the keyfile, so you'll need to back that up on paper too.

A word about using an internet-connected device to create your access plans. Most people aren't tech-savvy enough to securely erase a file from their devices (putting it in trash and emptying is not securely erasing). However if you are, feel free to create your inventory on your preferred device and then take the appropriate measures to securely delete all traces of it after you've printed it out. For most people, pen and paper is the best option.

Why is the printer and internet-connected device optional? If you're online and have easy access to a functional printer, you'll have the option to print out a few things to supplement your plan. The supplemental pages are nice to have but they're not necessary. However, don't make copies of your plan on your printer or scanner, as most devices today are wifi-connected and store data.

What kind of envelopes? For now, whatever envelopes you have are fine. You don't need anything special or fancy. If you don't have envelopes that's okay too; you could just fold your paper in half and try to seal it another way (like with a stapler). The idea is that you want to know if other people are peeking at your plans, so you'll want to store them in a tamper evident way. "Tamper evident" just means you'll know if someone read your plan because, for example, the envelope will be unsealed.

Why Primitive Technology?

You may be wondering why someone who is so pro-tech is suggesting you use such primitive technology. Why physical security instead of digital security, pen and paper instead of digital media, and lock boxes and tamper-evident bags instead of PGP, encryption, and digital signatures? The answer is simple. Because these primitive technologies work; they're simple, yet very effective. They can be used correctly by most people without security expertise.

Most people vastly overestimate their own skill with digital technology, especially before they've actually tried to apply it. But that doesn't even come close to how much they overestimate the skills of their heirs. Your heirs are much more likely to be able to recover your assets, and be able to understand the process of transferring your assets, if the process involves something familiar—like words on a piece of paper. Each electronic barrier you place in front of your loved ones reduces the likelihood they can recover your assets when the time comes.

I don't advocate primitive technology because I don't use digital security, quite the opposite. It's because I *do* use digital security: PGP, symmetric encryption, SSS, FIPS-certified storage devices, hardware multifactor, and I know how my heirs look at me when I start talking about these things—they are uninterested and clueless.

I use the methods recommended in this book for my personal estate planning; my clients use them too, and that's why I'm recommending them to you. In practice, they just work better. Do not mistake complexity for sophistication. When it comes to inheritance, robust, secure, and simple technology is best.

Select A Storage Location

The first step in making an access plan is to find a secure location to store it. When you're finished working on your plan, you'll put it into an envelope, seal it, and store it. In the Making it Better section, we'll discuss how to choose good storage locations and how many you should have. For now, use the safest location you've got immediate access to. If you don't have a safe or other secure location yet, you can designate a locking drawer in your office or a drawer in your bedroom as your storage spot: somewhere that's hidden but not too hidden. The right place is somewhere your loved ones will find the plan if something happens to you but not before. Remember, your plan isn't top secret; if someone finds it, that alone won't be enough to steal anything.

Use extra care with your backups and access codes — as they may be sufficient to steal your assets. Your wallet backups and access codes should be stored separately from your plan, if possible, in different, even more secure locations, preferably waterproof, fireproof, and access-controlled.

Now that you've got at least one storage location selected, it's time to move to the next step.

Selecting Helpers

Name helpers. If you don't do anything else, do this. Think about who your heirs should reach out to when they need help discovering and transferring these assets. Unless your heirs are just as interested and skilled at managing cryptoassets as you are, they will probably need help to actually access and securely transfer these assets. Remember, they will be in mourning, and everyone grieves differently. They probably won't be thinking clearly, and will appreciate having the option to contact people you trust to answer questions and make the process easier. Even if your heirs understand how to access your assets, they'll probably need help to learn how to hold them securely or liquidate them.

Let me be clear, I'm not suggesting you hire or designate a trusted third party. That is the antithesis of bitcoin and open blockchains. Instead of trusting a single third party, we're going to dilute the amount of trust required and distribute it amongst different people, and we're going to empower your heirs with information and educational resources, to help them educate and protect themselves.

Your Heirs Already Rely On A Helper—YOU

In order to build a plan that will actually do what you want it to—allow your heirs to access and transfer your cryptoassets—you need to be honest about their technical abilities. Many of my clients want to create plans that will allow their heirs to transfer assets without any help from a third party. I share that dream. But for now, it's just a dream for most people. Later, we'll discuss training your heirs; for now, we have to take people as they are, with the knowledge and experience they have today.

Take a minute and think honestly about the technical abilities your loved ones. How many of them are using password managers? Two-factor authentication on their email? When you videochat, do you sometimes only see the top of their head, or worse, straight up their nose? How many of them contact you when they're having computer issues? Are you holding cryptocurrencies for them?

Are *you* their helper? If so, you can assume that when you're gone there'll be a vacancy, a job opening. They will need someone to help them with their tech stuff, including your cryptoassets. And they *are* going to find someone to fill that open position. The question is, are they going to fill it with someone you trust or some random person they find online?

Or your cousin Billy who has a Coinbase account and thinks he's an expert on cryptoassets but actually has no idea what he's doing?

The best thing you can do for them is to give them options, and one of the most important options you can provide is someone to guide them through the process.

High-Value Transfers, High Stress

Think about how stressful it is for you when you're doing a "big" transaction. Big value transfers stress people out. Here's a little industry secret that people don't talk about: even the network and security experts among us stress out for a few seconds when we're transferring high values. Even in the traditional banking world, most people get nervous if they wire a lot of money to someone. They anxiously wait for confirmation that the transaction is complete, final. That stress is amplified in a cryptoasset setting, where there is no third party to intervene if something goes wrong. I'm sure you can identify with the feeling of not being sure if you got the address and amounts right (even though you triple checked them). Waiting, refreshing the page every few seconds to see a confirmation. Imagine the stress for your heirs, trying to access or transfer your legacy, which might be more value than they ever dreamed of. While grieving.

No matter how good your instructions are, it's unlikely they'll want to do it alone. They won't have the confidence to do it themselves. Which means they will probably start looking for help. They'll start Googling. Or posting on Reddit. Or going to a local "investment" meetup group. Is that where you want your heirs to find help?

From a practical perspective, it's better to provide your heirs with as many tools as possible, and that includes identifying people or organizations to help them if needed. Your heirs will find and use the tools they need, which may or may not include contacting helpers. Better that you decide who helps your heirs than rely on chance.

Let's Be Honest About the Technical Requirements

The other day I was taking a break from writing and decided to check out a couple of newer projects in the industry. After a bit of research, I decided to buy a few tokens of two of the projects. I traded some litecoin (LTC) for two different tokens, the transfers went through easily and quickly. Great. A few days later I learned something negative about one of the projects; it's far more centralized than I initially thought. So I decided to get rid of that token. I tried to trade it back to litecoin but I couldn't. Why not? Because the project was built as an ERC token on top of ethereum, which means I had to have ether (ETH) in my wallet in order to pay the gas (network fees) required to get rid of the token. At the time I didn't have ETH, so getting rid of the token required two transactions: first I had to buy ETH, then I could trade the token back to LTC.

I didn't need to own ETH to receive the token, because the exchange kindly paid the fee, behind the scenes. While that's nice, it set me up for an unexpected user experience when I tried to divest from the token. For me it wasn't a huge deal, because I realized what was happening. But imagine your heirs' reaction. Would they understand the error message "You need ETH to complete this transaction"? Would they think their wallet had a virus? Would they understand how to get ETH? Would they know how much to get? Would they give up and just decide it's not worth it? Would they look at the price of 1 ETH and think that's how much they needed to purchase in order to make the transaction go through?

This story illustrates two inheritance planning points. First, if you're holding tokens that use another blockchain as a platform, then consider holding some of the underlying blockchain's currency. Keep in mind, however, it's impossible to say how much because that's dependent on how many tokens you will have and what network fees will be in the future. Second, even well-designed systems can create confusing user experiences and unanticipated requirements.

Your heirs will need help to navigate this complex environment. No one can anticipate all of the possible scenarios, which is why selecting a helper is one of the first things we do.

Let's Be Honest About Your Ability To Write It All Down

For most people, keeping an instruction manual up-to-date is an impossible goal. Why? Because it requires you to fully understand the security and issuance models of every asset you own, identify your own assumptions, identify the knowledge gaps of every one of your heirs, and be able to explain all of that in a clear manner that they will understand. And if you get it wrong, or if they misunderstand, or if they simply miss a step, they could lose some or all of the assets. Permanently.

Moreover, this requires you to keep up with all of the technology upgrades for every cryptoasset you own, and then be able to explain the relevant ones to your heirs in writing. Think about your inner reluctance to begin this Get It Done plan and then think about the motivation needed to constantly update your instruction manual. That's an epic task. Basically you've just created an unpaid full-time job for yourself writing a tedious reference book for your heirs to read. Oh, and part of that book will probably need to be written on a device that is never connected to the internet—hope your handwriting is good!

In 2017, bitcoin, which is the slowest moving, most conservative network, saw a lot of change. For example, the introduction of multiple airdropped coins (which your heirs could miss out on if they don't have a competent trustworthy tech helper), a changed fee environment, a new address format, segregated witness, and lightning network deployment. Technical advancements and upgrades have been deployed in every cryptoasset network currently in existence. Some of these upgrades could affect token accessibility in the future. Again, you have to keep up with everything and be able to explain it to your heirs in a way they understand and can actually deal with. Rather than try to keep your training documents up to date, a more realistic plan is to ensure you have designated helpers who are already keeping up with the industry.

Identifying People To Help

As discussed previously, you'll want to identify people or organizations who can help your heirs and your estate with these assets, either by answering questions, or by actually helping with the transfers, if needed. In this section we'll discuss how to quickly choose helpers. In the Make it Better section, you'll learn to diversify the candidates and improve the selections. We'll also discuss educating your heirs and minimizing the need for helpers. But for now, if something happened to you tomorrow, your heirs would probably need help. So let's identify a few people or organizations who could help them.

Do not contact these people and ask them if they will agree to help your loved ones. We will discuss this issue later in the book, but for now it's enough to simply identify helper candidates.

Look through the contacts in your phone, email, and/or social media. Who do you go to when you have questions about this technology? Do you know any of those people personally? If so, do you trust them? If so, choose the three people you trust the most—ideally they won't know each other—and write down their names so you'll remember them later. Do not contact your prospective helpers at this point.

If you don't have three people like that then think of lawyers and/or organizations that might be able to oversee the process. They need to be reasonably tech-savvy and should have some training on cryptoassets. Lawyers are a good choice, if they're trustworthy (which you shouldn't assume from the title alone). Lawyers will take these issues seriously, and most have malpractice insurance which might provide your heirs with some recourse if the lawyer screws up badly. If you're looking for a lawyer who has some knowledge about these issues, you may be able to find one through our Empowered Law [https://empoweredlaw.com] legal directory.

Beyond finding "expert help," you might want to add a trusted independent person to your list. Someone who might not really know cryptoassets but is extremely trustworthy, not afraid to ask questions, and if possible, reasonably tech-savvy.

There are incredible people doing incredible work in these industries. They are tech-savvy and many have made a name for themselves; they're famous in the cryptoasset industry. You might watch them on YouTube or listen to their podcasts. But just because they have lots of followers or a weekly show, don't assume they are trustworthy or the kind of people you should name as your tech helpers. Even if they are the most trustworthy people on earth, they may not actually have the time or ability to help when the time comes. So try to stick with people you know personally, or established companies that actually offer these types of services.

If you have an internet-connected device and printer available right now, then take a minute to print a photo of each of your helpers; this will help your heirs validate their identities. Don't worry if you don't have immediate access to a printer. Just move on to the next step.

Begin Writing A Letter To Your Loved Ones

Now it's time to begin writing your letter. This letter will serve as your primary cryptoasset access plan and it will contain all of the information they'll need to have in order to access your assets, when the time comes. You're free to use any wording you like, however if you're not sure where to begin, you might start with the sample language below. You could also use the tempate in the appendix Letter to Loved One Template.

Example 1. Sample Letter

> *Dear loved ones,*
>
> *Today is (insert date, including the year) and I'm writing this letter to let you know that I have cryptocurrencies and cryptoassets that may be worth something. Please read this letter carefully and completely before taking any action. These assets are different from other assets—once these assets are transferred, there's no way for you to get them back.*
>
> *Below is a list of people I trust to answer questions and help you through the process of finding and transferring these assets. Contact the people listed; do not trust only one person to handle this process. Watch all helpers very carefully, even the people listed here. Anyone can make mistakes, so make sure you understand what they are doing as best you can, and don't be afraid to ask questions and verify answers yourselves.*
>
> *People who can help answer questions and guide you through this process are:*

Write down the names and basic contact details for your helpers. Include more than one way to contact each helper, for example by email, Twitter, Keybase, or Telegram. If you have printed a photo of one or more of the helpers, write the name of the person and their contact info on the photo.

Now that you've identified who can help, let's look at what they can help with.

Building Your Quick Asset Inventory

In this section you'll tell your heirs what to look for and where to look for it. You'll be listing your access devices, the software you use, the location of your wallet backups, and the location of any other information required to access your assets, like passwords or passphrases. The reason you're listing your access devices is to ensure your loved ones don't accidentally give away or destroy devices that have copies of your keys or passwords before they move the corresponding assets to new accounts.

For legal and privacy reasons, we don't want to tie accessing your cryptoassets to accessing everything on your laptop, phone, or desktop devices. Instead, we'll rely on your backups and your plan to provide the necessary access information. For these reasons, you must make sure you have at least one full and complete backup of each key, seed, and necessary access code. I'm using the term *access code* to mean anything that is not key or seed material but is required in order to provide access to those keys or seeds, such as an encryption password.

When in doubt, back it up! If you're not sure whether your password, PIN, or code is required for access, back it up. If your heirs don't need it, they probably won't use it. But if they do and they don't have it, they might be locked out forever.

If you've been playing with cryptoassets for a while, chances are you have hardware and software devices you haven't actually used yet. Don't worry about those for now, ignore them. You may decide to use them later, but for now we'll stay focused on your holdings as-is.

Don't let yourself get sidetracked. Taking even the most basic inventory of your cryptoassets will probably lead to some "I should" moments. You'll think, I should consolidate these assets, or, I should sell that and buy this. Or, I don't really like this wallet, maybe now is a good time to move to that hardware wallet in my drawer that I haven't used yet. **Do not start moving assets now**. Once you start looking at prices and transaction fees, and get into the mindset of making changes to your holdings, you'll abandon the inventory process and won't actually take the steps necessary to protect your heirs. It's happened to many of my clients, and it even happened to me when I started my first cryptoasset estate plan. Instead of getting side-tracked, take 15 seconds to add a generic item to your calendar or your to-do list, like "consolidate cryptoassets" — and immediately return to your inventory task. Remember, no one is going to judge your cryptoasset allocation. Your inventory isn't a reflection of your ability to understand cryptocurrencies or your market prowess. It's simply a snapshot of your current assets. It doesn't need to be perfect; right now it just needs to be as complete as possible.

Phone

Start with your phone, simply because it's nearby. If you use your phone to access any cryptocurrencies or cryptoassets then you'll include those in your inventory. Start by identifying the make and model of your phone, then list all of the apps you use to access these assets. Then you'll verify that you have a backup of the wallet, and that you've backed up any necessary access codes. The easiest way to do this is to create a list or a table.

Example 2. Continue Your Letter, Add An Inventory And Device List

Below is a list of devices and software I use to access these assets. Please put all of these devices away, under lock and key; store them securely until the assets have successfully been transferred to my heirs. Do not let anyone access them without supervision.

Cryptoasset Inventory

Device	Software	Asset(s)	Backup(s)	Password(s)	Notes
Nexus 6	Samourai	BTC	Aunt MK & Best Friend	LockBox & BankVault	Insert helpful notes here

Backing Up Your Wallet

Now it's time to make sure you have backed up your wallets. Most wallets use the industry-standard term "backup" to allow you to make a copy of your private keys. With a copy of the private key (and passwords or passphrases, if necessary) you can use any compatible software to "restore" your wallet. "Restore" is just another word for "recreate."

Typically your heirs will need to recreate your wallet on another device, possibly using different software, in order to access your cryptoassets. Your heirs shouldn't need to access your phone or computer to claim your assets. This is one of the reasons backups are so important to estate planning. It's also why you should be very careful with your wallet backups: you don't want to lose them, and you don't want others to have them before you want them to.

Three quick points about backups. First, before you start making backups, be sure your location is reasonably private. Cover windows, close doors, look around the room for cameras, cover the cameras on your computer and phone, and make sure you're alone. Second, do not get cute. Do not create your own encryption scheme where you mix your backups with other words or data, and/or cut them into pieces and scatter them around like birdseed. Doing that actually *reduces* security and increases the likelihood that you'll lose the funds. We'll discuss these issues in more detail later. For now, just create a complete backup, intact, following your wallet's instructions. If your wallet instructs you that you need to back up your password, passphrase, or PIN, do that too, each on a separate piece of paper. Final point here: if you're backing up on paper at the top of the sheet, write the name of the software or hardware you're backing up. For example: Electrum, Mist Keystore Password, or Trezor Model T. This information might be crucial for helpers, especially if your software uses non-standard key generation methods.

Sample Backup Card

Below is an example of a backup card for mnemonic seeds that many of my clients choose to use. Each line is intended to provide relevant and important information to you and your heirs about the assets. Completing all of the information requested is optional, and your choice to do so will depend greatly upon your specific situation. This should be obvious, but just to be sure there's no confusion — do not actually use the sample seed for your assets.

ASSET ACCESS CODE: do not discard

Software: Copay
Date: March 2018 Additional Code Req: (Y)/ N
Assets: BTC, BCH
Notes: Part of 2-of-3 multisig w Alice & Bob

1	first	13	drip
2	harvest	14	thumb
3	walk	15	resemble
4	fancy	16	vendor
5	follow	17	error
6	crazy	18	path
7	day	19	trade
8	train	20	impact
9	purity	21	verb
10	glide	22	later
11	grape	23	club
12	victory	24	rug

Figure 2. Sample Seed Backup - DO NOT USE THIS SEED

Start by naming the software you use with the seed. Add the date, because it's helpful to know when the backup was made, as it could give clues to undeclared assets. Note whether or not an additional code is needed to access the assets, and then list the trading symbol or name of all of the assets being backed up with this seed. Use the notes section for any extra information you want to add; for example, "This key is part of a 2-of-3 multisig with Alice and Bob." You can find a blank seed backup template in the appendix of this book at Template Seed.

Sample Access Code Card

Below is an example of the access cards some of my clients use to back up their asset encryption passwords or passphrases. Again, how much detail you choose to include is highly dependent upon your situation. Because this book is about inheritance, and the primary goal is access for your heirs, if you're not sure what to include, err on the side of more information. If you're using a wallet with a keystore file, use this form to backup the encryption password and note that the keystore is stored on a USB, CD, or other electronic media. This should be obvious, but just to be sure there's no confusion — do not actually use the sample code for your assets.

ASSET ACCESS CODE: do not discard

Software: myetherwallet.com

Date: March 2018 Additional Code Req: Y / N

Assets: ETH, ETC, ZEC

Notes: use with Trezor seed

Password/Passphrase:

aardvarks play saxophones while

baking fuzzy butter cones

Figure 3. Sample Access Code Backup - DO NOT USE THIS CODE

You can find a blank seed backup template in the appendix of this book at Template Access Code.

Computers And Laptops

Now that you've finished with your phone, get your computer or laptop and return to your letter. What software do you use to access your assets on those devices? Are you also using a hardware wallet to interface with the software? List the make and model of the computer(s) you regularly use to access your assets, and the software you use to do so. Be sure to list all of the exchanges you use.

Example 3. Continue Your Letter, Add Assets You Access With Your Computer

> *I use my 2016 MacBook Pro 13" Retina to access my Cryptokitty using Metamask. The backup and password for Metamask is located: (insert general location, not specific address)*
>
> *I also use my laptop to access my Poloniex exchange account. I currently have some USD, LSK, and ZEC held there.*
>
> *Contact the exchange to get access to my account (because it's illegal in many places for you to use my login info to access them after I'm gone) OR you can find my login information at: (insert general location, not specific address)*

For custodial accounts, or those controlled by third parties, you will need to consider whether or not you want to give your heirs your login credentials. You can learn more about the benefits and pitfalls in Felonious Heirs. If you want your heirs to use your login credentials and you have 2FA enabled, you will also have to provide login information for your phone or mention your 2FA device in order to enable access.

If you have an automatic buy set up on an exchange with your traditional banking institution, consider mentioning that here so your heirs can take action more quickly.

Example 4. Continue Your Letter, Add Cold Storage

> *I also have assets in cold storage on both my Trezor Model T device and Ledger Nano S device. On the Ledger, I store BTC and BCH and use Ledger's software to access it. On the Trezor, I store ETH, ETC, and ZEC using myetherwallet.com.*
>
> *The backups for each device are located: (insert general location, not specific address)*
>
> *I use an advanced security passphrase for my Trezor device, the backup for the passphrase is located: (insert general location, not specific address)*

If you use advanced security features, like a passphrase on your BIP39-compliant seed, or Shamir's Secret Sharing, or you've encrypted your keys or devices, be sure to have that access information available for your heirs and stored securely (but not in the same place as you store your backups). If you don't understand what I just said, don't worry about it; it doesn't apply to you right now, and we will discuss it later in the book.

If you have an internet-connected device and printer available right now, take a minute to print a photo of each of your devices—especially if you're using hardware wallets. Just use the stock photos on the manufacturer websites. This will help your heirs recognize these devices and secure them appropriately. After you've printed the photo, write the name of the device on the photo. Again, if you don't have immediate access to a printer just move to the next step.

Next, think about devices and assets you haven't mentioned

yet. In the Make It Better section, we'll go over a huge list of assets to jog your memory. For now, just mention the items that immediately come to memory, like paper wallets, Cryptosteel, or Opendime.

Example 5. Continue Your Letter, Add Other Cryptoassets You Haven't Yet Mentioned

> *I also have paper wallets and a Cryptosteel stored at: (insert general location, not specific address)*

Remember, you don't need to explain what these devices are at this point. They can look online and/or ask the helpers you've designated.

All you have to do now is close your letter. If you already have a will, testament, or trust, you should mention it here. If you have an estate planning lawyer or accountant you haven't already mentioned, you should mention them here too. You might want to add something about why you got into these assets and remind them that you love them. Some people even include a joke or a poem; remember they'll probably be emotional reading the letter, so you might want to try to make them smile if that's your style. This isn't a legal document; you don't need to sign it or have it notarized.

Example 6. Finish Your Letter, Add Legal Information and Your Closing

> *You will find a copy of my will, dated April 17, 2018, in my file cabinet. My lawyer, Bob Laublaw, of Chicago, Illinois also has a copy of it. Remember, I love you very much.*

Wrapping It Up

Let's be sure your basic access plan includes as much as possible. Did you include information for:

- ☐ Helpers: names, contact information, photos if possible
- ☐ Devices: your phone, computer, hardware wallets, paper wallets
- ☐ Wallets: all of the software you use to access your assets
- ☐ Assets: include a list of assets
- ☐ Exchanges: be sure you've listed all of the exchanges that are holding funds for you
- ☐ Access: the information they'll need to find your storage locations and all necessary access codes

If not, go back and add the missing pieces.

Make A Copy

Consider making another handwritten copy of the letter to give to a loved one, lawyer, or to store in an alternative location. This step is optional but it will increase the chances your loved ones will find the letter in time.

Educate Your Heirs

Some people really want their heirs to understand the process before they'll need it. They trust their heirs completely, share everything with them, and are not interested in keeping anything from them, including how to access their cryptoassets. If that sounds like you, then consider having your primary heir (perhaps your life partner) write a copy of the letter for you. As they write things down they'll have questions; it's a good time to discover those questions and clarify. It can also be a really empowering experience for your heir. That said, you cannot un-share this information; please think carefully and critically before you do so.

Store Your Letter and Backups

Your letter is now ready to be stored. Place the letter in an envelope. Seal the envelope and sign the seal. The reason you're signing the sealed part is so that you or your heirs will know if someone tampers with the envelope, for example to read the letter or replace it with a new one. Then place the envelope(s) in the storage spot(s) you selected earlier.

If you've backed up your wallets, passphrases, or other necessary access information during this phase, be sure to store them securely, separately from your plan if possible. If you have not yet backed up your wallets, **do it now**! Common secure storage options for home include a locking file cabinet (should not be your only copy), personal fireproof/waterproof safe, or a gun safe. Common secure locations outside the home include a bank safe deposit box, a vault in your hometown, a vault outside of your county or state, a vault overseas, your attorney's office, your accountant's office, a relative's fireproof/waterproof safe, or a fireproof/waterproof safe at your office.

If you are not storing your wallet backups in a fireproof, waterproof, locked location, you should deal with that immediately. You can order a safe on Amazon, pick one up at your local office supply store, or rent a safe deposit box. If you have no other waterproof option, put them in a waterproof bag; in a pinch a plastic freezer bag can work. You should also consider buying numbered opaque storage bags for easy tracking; they're available on Amazon. You should also plan to diversify your storage locations as soon as possible. What you store in each location depends on how likely you think it is that someone will access that location without your consent. We'll explore these ideas in greater detail later in this book.

Schedule Time to Make It Better

Before you run out to celebrate, let's quickly set aside time to make your plan even better. Get your calendar, or open your calendar app, choose a day and designate two to three hours to start working on the improvement phase.

Celebrate!

Congratulations: you've completed part one; you got it done! Hopefully you're feeling a sense of accomplishment. I also hope that it was easier and faster than you thought it would be.

Make It Better

Now that you have a basic plan in place, let's look for ways to improve your plan, to make it work even better for you and your loved ones without making it a constant hassle. There is no cookie-cutter solution with inheritance planning. Everyone is different, and while most people share some concerns—like keeping their own private keys private—some people are quite comfortable sharing their keys with their spouse, parents, and other loved ones. While I shudder at the thought of sharing keys, ultimately it's not my decision to make. It's yours and yours alone.

In this section, we'll start by discussing what makes a good access plan, and common risks. Then you'll evaluate your current plan, identify ways to improve it, and finally get to work on improvements. Hopefully it hasn't been too long since you've completed your Get It Done plan.

Access Plan Dos and Don'ts

There are some simple rules to remember as you're revising your Get It Done plan.

Do Decide Who You Want to Control the Access Process

In a typical estate, one person is selected to oversee the process of asset distribution, they're often referred to as a personal representative or executor and the process is often referred to as estate administration. The personal representative controls all of the assets, on behalf of the estate, and transfers them to the rightful heirs as indicated by the will, or by the court if there is no will. Ideally, the executor is not one of the beneficiaries of the estate; ideally they are an independent third party. Because your cryptoassets *will* be part of your overall estate, one person will probably be in charge of distributing them just like any other asset. However, you can choose how many people or organizations need to come together to enable the executor to do that job. For example, you could design a plan where the executor would need to access two separate bank vaults or safety deposit boxes in order to access your backups and other necessary access materials. Or you could design a plan where a number of people, who do not know one another, have to come together in order to provide the necessary materials. As you build your backup plan and select your storage locations, think about who you want to be involved in the access process.

Do Include All of the Information Your Heirs Will Need

An access plan is intended to be a roadmap for your authorized heirs to find and access your assets when the time comes. A good access plan will include all of the relevant information your heirs will need in order to gain access to your assets, if they can find and access your secure storage locations. A good access plan includes warnings for your heirs about opportunities for theft, and provides resources to help them learn about the assets.

Do Write The Plan For Your Heirs, Not For You

Write the access plan for your heirs, in clear language they will be able to understand, and make sure it recognizes their technical abilities as they are, not as you want them to be. It can be written in prose, or contain tables and screen shots if you like. It should be a reasonable length; two to five pages is sufficient for most plans. Your plan should be stored somewhere that your heirs will easily and quickly find it after you pass away.

Don't Make Your Access Plan a Top Secret Document

Your access plan is not a confidential document. If someone sees your plan, they shouldn't be able to steal your assets. It shouldn't contain key materials; those should be held in secure storage locations. Don't store the only copy of your access plan in a safe deposit box or bank vault, because accessing those after death requires legal process and usually significant time. Many people choose to store their access plans in their home safe or with their lawyers. That said, you shouldn't leave it laying around either. Don't provide an opportunity for someone to gain personal information about you that they don't need.

Don't Make Your Access Plan a HODL Plan For Your Heirs

We are not designing a holding plan for your heirs. In order to hold cryptoassets securely, your heirs will need to learn about them and make decisions about what will work for *them*, not what you want to work for them. You cannot and should not try to design a long-term storage solution for them.

When you move into a new home, you should always change the locks because you don't know who has copies of the keys and when they might decide to use them. The same is true for cryptoassets. This is why your heirs should transfer their inherited cryptoassets to keys they generate, own, and fully control.

Cryptoasset owners must understand and trust the people and the processes involved in key generation, division, and storage. If you really want them to be able to inherit the assets directly, you should work with your heirs to help them design their own long-term storage plan while you are alive. Regardless, their long-term storage plan should not be part of the access plan for your assets.

Don’t Make Your Access Plan Your Will Or Testament

An access plan is not a will or testament and should not include directions on how to distribute these assets to your heirs. If you want to have control over distributions you’ll need to follow the law in your jurisdiction to formally write a will or testament, or put your assets into a trust. All of these things are covered later in the Make it Legal section of the book. For now it’s enough to know that you should not use the access plan to devise these assets.

Now that you understand the basic dos and don’ts, it’s time to look at your current plan and Make it Better!

Be SURE

The most challenging parts of inheritance planning for cryptoassets are making decisions about:

- where and how to store your wallet backups and access codes,
- where and how to store your inheritance plans,
- if and how to share information with your loved ones,
- when to share that information, and
- what information you should share.

These are difficult choices for everyone. Most people don’t have the necessary expertise in security to be able to evaluate and balance risk, and that makes them feel vulnerable, unsure of what to do. The most common reaction to that vulnerability is to simply give up, to do nothing. The problem with that strategy is that it almost always results in your heirs losing those assets.

Balancing security, usability, resiliency, and efficiency is possible. In the next few sections, I will help you evaluate your risk, understand the threats you face, and make *sure* your plan is well-balanced and addresses the hazards: Securely, Usably, Resiliently, and Efficiently (SURE).

We use the SURE acronym to remind us of the relative importance of each of the elements of our plan. A properly balanced plan is not one where all things are treated equally; some things are more important than others. And while there will be some differences in each person's unique circumstances, most plans will follow the SURE prioritization. What that means is we prioritize security above the other elements, but not to their exclusion.

First, our assets need to be secure against theft. Second, the plan must be usable for our heirs. Our heirs and helpers will already be at a disadvantage because they didn't write the plan. They're trying to figure it out from limited information. We need to make the plan as simple to use as possible, without sacrificing too much security. Third, our plan must be resilient. We shouldn't lose everything because a single piece of paper got soggy, lost, or accidentally discarded. Failure is an option; in fact it is almost a certainty when there isn't a well-balanced plan. Our plan should survive several minor failures without the loss of assets. Finally, our plan should be efficient. We're trying to avoid unnecessary cost and complexity. We'd like to avoid a 20-year treasure hunt probate, spanning three continents, and 150 hidden stashes; your access plan is not the plot of *The Da Vinci Code*. For the majority of estate plans, three or four secure locations and two or three trusted people is sufficient to be SURE.

Set a Goal

You'll begin this process by identifying the balance of SURE you'd like for your plan. Using the table below write down a number from 1 to 10 that conveys the relative priority of this element, with 1 being unimportant and 10 being most important. At this stage we're not doing any analysis. You're using your intuition based on your unique understanding of your circumstances, environment, threats, and fears.

Table 1. Identify Your SURE Goal

	Security	Usability	Resiliency	Efficiency
My Goal				

Look at the goals you set above. Unless you have truly exceptional circumstances, S should have the highest number and the rest should be slightly lower. You should not have all 10s because all things are not equal, and a plan like that is not achievable. You should not have wildly diverging numbers, like one 10 and three 2s. A gentle slope from left to right is what most people need. Feel free to change your goals until they feel like they match your risks, your values, and your circumstances.

Now that you've set some goals, let's talk about how to make your current access plan better. Then we'll re-evaluate your plan in terms of SURE.

Better Asset Inventory

In the Get It Done section, you quickly listed the assets that mattered most. Now we're going to spend some thinking about your assets, what you have, what you might not remember you have, and then we'll consider consolidation and secure backups.

The purpose of a comprehensive inventory is twofold: first, you'll be able to better manage your assets while you're alive (because you'll know what you have), and second, it will help your heirs locate those assets when the time comes. It's best if you leave the asset inventory for your heirs, instead of destroying it. If your heirs know what you have, it will be much easier for them to access it. Over the past year my thinking in this area has evolved. I used to think that an asset inventory would hurt more than it would help because it's unlikely your inventory will be accurate at the time you pass, especially if you're an active trader. However, without an inventory, the recovery of assets becomes much more complex and time consuming, because wallet software isn't always standardized and wallet apps add and remove token support at a whim. If you're using closed-source wallet software, that asset discovery process becomes even more difficult. I now believe the benefits of providing a comprehensive inventory outweigh the risks, and most of my clients choose to make the comprehensive asset inventory part of their access plan.

To try to ensure we don't miss anything, we're going to review the assets in two different ways: first you'll try to list all the cryptocurrencies and tokens you own and where they're held. Then you'll review a list of common assets and currencies to help jog your memory. If you're newer or haven't played around with a bunch of different assets, this part of the planning will be finished quickly. If, however, you're like me and like to try out new wallets and new tokens, this might take some time. It will be worth it.

First, start with the inventory from your Get It Done plan. As a reminder, it might look like this:

Table 2. Cryptoasset Inventory

Device	Software	Asset(s)	Backup	Password	Backup2	Password
Nexus 6	Samourai	BTC	S1	S2	S3	S4
Nexus 6	Copay	BTC, BCH	S2	-	S4	-
Laptop	Poloniex	USD, LSK, ZEC	-	-	-	-
Laptop	Metamask	Cryptokitty	S3	S4	S1	S2
Laptop	Copay/ Trezor	BTC, BCH	S1	S2	S3	S4
Laptop	Ledger/ Ledger	BTC, BCH	S2	-	S4	-
Paper Wallet	-	BTC	S2	S4	-	-

Now try to add to the list. Have you ever accepted cryptocurrency or tokens in exchange for services? How did you get your first tokens? Do you still have that wallet? Search your computer for old files. Search your old hard drives for old files. What about your old phone? Do you still have it laying around? Any chance it still has a few satoshis or weis on it? Now is the time to check. Take a look through your password manager files (if you're not yet using a password manager you need to be; add it to your to-do list for this week!).

If you want to learn more about using a password manager, feel free to read my free article about how to use password managers [https://empoweredlaw.com/articles/articles-2/].

Have you ever owned: colored coins, open assets, gaming tokens, artist tokens (like TatianaCoin)? Have you used LocalBitcoins, or do you have any loose change left over in an exchange or old wallet? Have you received cryptocurrency tips, for example from Tip Jar? How about browser-based wallets for minor spending, like Brave? Do you have physical representations of digital assets, like Casascius coins or paper wallets? What about Opendime?

Once you've exhausted your memory and the list above, take a look at the List of Cryptocurrencies page on Wikipedia [https://en.wikipedia.org/wiki/List_of_cryptocurrencies], Coincap [http://coincap.io/] and for tokens CoinMarket Cap [https://coinmarketcap.com/tokens/views/all/]. As of April 2018, there are more than 1,500 tokens listed on CoinMarket Cap. Be careful, though: you might get sidetracked looking at all the projects you didn't even know existed! Now you know why we didn't start with this—there's always so much to learn.

Update your inventory sheet as you go.

Consolidating and Evaluating Assets

Now that you have a complete list of what you've got, take a few minutes to think about your holdings (or hodlings!). Are there any tokens you really don't want anymore? If so, consider selling now or trading them for a token you do want. If you don't want to use a traditional exchange, you can use Shapeshift. If you want to keep them, consider consolidating them onto one or two multi-token multi-currency seeds. Depending on network congestion and fees, now might be a good time to do that. If not, make a note on your calendar to do it later.

In some jurisdictions, trading one asset token for a different asset token is a taxable event. You may want to consider the tax implications before you begin to consolidate.

When I reviewed my password manager files I found an old Counterparty wallet I hadn't looked at for 3 years and guess what — there was 0.25 bitcoin sitting in it! At the time, that was worth around $75. The value today is much more than that. Needless to say that was a happy find! And I decided to move it right away to a new wallet that I use regularly and have properly backed up.

As you think about consolidating your assets, you might also want to think about improving the security of your cryptoasset holdings by implementing a "separation of powers" strategy.

Separation Of Powers

One of the most important questions we need to consider when building our access plans is how many people need to come together to steal the assets. At this point ignore inheritance allocation for a moment and ask yourself the following question: **who do I want to have access to my complete backups?** Put another way, who do you want to be able to have full power over these assets? Some people say, my spouse, my best friend, or my attorney. Those people are perfectly comfortable with a single person having complete control over the assets. However, if you've selected one person, remember they will be able to misappropriate the assets, or even hold them hostage, in order to get things they want from the other heirs. This happens with non-crypto assets all the time—someone with the keys to the house goes in and takes the valuables they say were promised to them, or they think they deserve. Then everyone else is left with a few terrible alternatives: overlook it, go to court, or attempt to make good in some other way. All of these lead to unneeded cost, drama, and years or decades of feuding. Therefore, instead of trusting a single person with the keys, it's usually a better choice to divide the key material into pieces and require more than one person to come together to unlock the cryptoassets.

If you're a newbie, or not particularly tech-savvy, the separation of powers discussion that follows may be a bit techy. Don't worry, you don't need to do anything about it today, and if you decide it's not for you then skip to the Physically Dividing Your Key section for some fun.

In the land of cryptoassets, the person with the key holds all the power. From a technology and network perspective, the keyholder has absolute power to transfer, sell or spend the assets. In some circumstances, a partial key holder can prevent others from transferring, selling, or spending (by refusing to cooperate). Within these networks, keyholders' rights don't come from law or legal jurisdictions, they come from the rules of the network.

Legal rights are distinct from network rights, and while they might align, they don't have to. This is a good thing. Remember, it would be nearly impossible to discern legal rights in real time on a global, fluid network, and it would require location and jurisdiction to be imposed on every transaction.

The network rules (and only the network rules) govern the powers within the network. This matters for estate planning because, if you give your cryptoasset keys to anyone, they will have absolute power over the assets, even if legally they don't have rights to them. So how do we solve this problem?

Simple. We don't give the keys to one person; we divide the keys, we divide control! As you might imagine, there are some good methods for dividing control and there are some terrible ones. I've talked to hundreds of people around the world about how they deal with this issue, read thousands of comments on Reddit, and helped my clients design plans that work for their loved ones and their unique situations. Over the years I've seen four primary methods that the overwhelming majority of people use to accomplish division of control for their cryptoassets.

These four methods are (1) multisignature, (2) backup + passphrase (BIP38 or BIP39), (3) Shamir's Secret Sharing, (4) physically dividing up the keys by cutting up the paper they're written on (or something similar). The first three technology-based methods can all be good choices, depending on your needs, however the fourth method is terrible. Remember our acronym, SURE? Physically dividing keys is a terrible option because it isn't secure, usable, resilient, or efficient. While a full technical analysis of each method is beyond the scope of this book, there are a few things you should understand about each of these methods in order to decide which is best for you.

For each of the methods, we're going to look at how SURE they are: we'll be looking at security, usability, resiliency and efficiency. "Security" means, in an ideal implementation, how secure the overall scheme is from the perspective of someone who has access to one piece of the key. Put another way, if a bad actor gets ahold of one of your backups or devices, what will it take for them to steal your assets? "Usability" means how easy it is for non-technical people to use the method securely. "Resiliency" means how many points of failure will result in a catastrophic loss. Resiliency is a vital component of good inheritance planning. And we want to do all this efficiently. Given enough time, we expect one or more parts of our plan to fail ("fail" meaning "does not turn out as planned") because of human error, electronic media error, natural disaster, changes in technology, changes in our relationships, and lots of other reasons. One or two of these failures should not result in total loss.

Multisignature

Multisignature is exactly what it sounds like: more than one key signature is needed to transfer the asset. The most common multisignature scheme is 2-of-3, where three signatures are designated as "authorized" and any two of those authorized signatures are required for a transfer. For example, if Jessica, Erica, and Lucy wanted to save for a vacation together, they could create a multisignature address where anyone can send funds to the address, but in order to take funds out, at least two of them must sign the transaction. Although the 2-of-3 configuration is the most common, you can actually create multisignature schemes to require M-of-N signatures, where M is the number of signatures required and N is the total number of authorized signers. In bitcoin, N is currently limited to 15 signers. Although the bitcoin protocol has a multisignature-specific implementation see Bitcoin Improvement Proposal 11, BIP11 [https://github.com/bitcoin/bips/blob/master/bip-0011.mediawiki] it is usually implemented using Pay-to-Script-Hash (P2SH); see BIP16 [https://github.com/bitcoin/bips/blob/master/bip-0016.mediawiki].

Multisignature controls can be implemented at the blockchain or consensus level, which is the most secure, or by a particular application, which is much less secure. How can you tell the difference? One way to tell is to ask this question: Can I change the number of signers or the authorization keys without a publicly-visible transaction on the blockchain? If the answer is yes, then you're using a less secure application-policy–based service. If this is the case, be aware the authors of the application actually control the asset, and none of the benefits I'm about to explain applies to those services.

> Often wallets with a "vault" feature will implement policy controls like spending limits, delayed transactions, and extra approvals. Policy controls aren't bad, in fact they're necessary whenever you have more than one person with signing authority. However, from a security perspective they don't replace or even come close to equaling the security provided by blockchain-based or consensus-based controls.

There are many benefits to using multisignature as a method of dividing control. First, it offers protocol-level security, meaning that no-one can bypass the controls. Anyone who wants to spend the funds locked in a multisignature address must have access to more than one key. There is no courtroom work-around for the consensus rules: no keys, no access. Period. If we wanted to steal the vacation fund, we would need to gain access to Jessica's and Erica's keys, or Jessica's and Lucy's keys, or Erica's and Lucy's keys. Unless they all live together and keep their keys together, compromising more than one of their keys is much more difficult than getting ahold of just one. Additionally, this configuration gives each person a backup. If Jessica's keys are lost or stolen, she can work with Erica and Lucy to create a new multisignature address and move all the vacation funds into the new address. This is assuming, of course, that Erica and Lucy decide to do so.

Back to the pros of multisignature. Another major benefit is that it's standards-based—although not all implementations are. Standards-based means that you're not just using a single company's idea of what multisignature should be, but an industry standard. That's good because if the company goes under or you want to change software, it's much easier. This is another reason that, from a consumer standpoint, open source software is so much better; even if the company or individual who wrote it disappears, the software will likely remain available, and quite possibly others will step up to support it. Today there are multiple implementations of multisignature, and they're fairly easy for non-technical people to use. Also, multisignature doesn't have to mean multiple people; one person could have all three keys. Finally, from a security perspective each key is independent; having one key does not give you any clues as to the identity of the others.

If you have never spent or sent funds from a multisignature bitcoin address, then the compromise of one key offers no clues as to the others. If you have spent funds from the address, then the redeem script (list of authorized signing keys) is published on the blockchain, so if one of the keys is compromised it can be used to identify the multisignature redeem script (and address) it is part of. In the future, technologies like Merkleized Abstract Syntax Trees (MAST), Taproot and many others will likely be used to improve privacy for P2SH transactions.

One big disadvantage of a multisignature scheme is that it's not widely available for all coins and tokens. Today, you can use multisignature in litecoin, monero, ripple, bitcoin and its forked lineage (bitcoin cash, etc). While multisignature ethereum smart contracts are technically available, they have not proven to be safe to store large amounts of value. In November 2017, $150 million USD worth of value accidentally became instantly and permanently locked in a variety of multisignature smart contracts by a single user [https://bitcoinmagazine.com/articles/de-briefing-ethereums-parity-predicament-whats-next/].

Multisignature is a great solution for some cryptocurrency owners. Many of the multisignature wallets available today have streamlined the address creation and spending processes — they've built easy to use interfaces with a built-in transaction approval and signing workflow. Today it's fairly easy for a non-technical person to implement and use. Additionally, it's easy to build resiliency into a multisignature system, either by widening the gap between M and N, for example 5-of-12, or by making extra copies of the keys themselves and storing them carefully. As I'll remind you later in the storage section, obviously there is no point to creating a multisignature address if you store all of the signing keys in one location. From a security perspective, requiring multiple keys puts a barrier in place if one of your keys is compromised. Even if one of your keys ends up in the hands of a bad actor, without at least one other key they can't do anything.

In terms of estate planning, there is another benefit to using multisignature to build an internally-redundant solution—accountability. Because each key is generated independently, a signed transaction reveals which keys have signed. If you see that a key has signed, you can draw the conclusion that either the original owner of the key signed the transaction, or they have lost control over their key.

Key or Seed, Plus Passphrase

The second common method of dividing control is to use a key or seed and a passphrase to lock your cryptoassets. Unlike multisignature, which invokes a blockchain- or consensus-based rule, this method is enforced by cryptography. In bitcoin, there are two different ways to implement this method, and the implementation depends on the type of key you're starting with. For you nerds, BIP38 applies to a single private key, while BIP39 applies to seeds. Seeds are used to build HD ("hierarchical deterministic") wallets. In HD wallets one set of words is used to generate as many unique addresses as you need, on demand, and that set of words can recreate all of the keys used in that wallet. Explaining in greater detail how HD works is beyond the scope of this book, but don't worry because you really don't need to understand it for estate planning purposes. And, although there are significant technological differences BIP38 and BIP39, for our purposes the differences don't really matter. What matters to us is that the passphrase acts as a cryptographically-enforced second factor. It's required to recreate your key material, and without it no-one can unlock funds or transfer assets even if they have a complete copy of your wallet backup. If someone tries to use the seed or key alone, without the passphrase, they'll simply see a wallet with no funds.

Notice that this second factor is not the same as 2FA (two-factor authentication), like when you use your phone and a password to access a website. It doesn't involve Authy, Authenticator, or, worse, an SMS. To learn more about 2FA see Bitcoin Security Made Easy [https://medium.com/@pamelawjd/bitcoin-security-made-easy-using-2-factor-authentication-92c3943471fa].

It's not the same as creating a PIN to approve spending on your wallet; those are all policy-based controls. A passphrase is not generated by an external company or device; you create your own passphrase. It's also not the same as a password that you use to access your online accounts. Passwords can be reset or recovered; lost or forgotten passphrases cannot be recovered.

In order for you to evaluate if this is the right method for you, you need to understand a bit about how it works behind the scenes. I'll use BIP39 for this discussion, because it offers multicurrency support and has become the standard for many of the most popular wallets today. Cryptographers among you, please forgive me for oversimplifying this explanation. In a BIP39 implementation, the passphrase effectively serves as a pointer or a location beacon for where the keys to unlock your assets are stored. If assets are protected by a seed plus passphrase, the seed alone will not provide access to the assets, because the seed alone will point to a different, likely empty, location. In BIP39 your seed is not encrypted with your passphrase; rather, your passphrase is included as entropy in the stretching algorithm that produces your keys. This matters because there is no error checking in a passphrase implementation, meaning if you type the passphrase incorrectly no error will appear. In 2016, I delivered a Bitcoin Security presentation at a Bitcoin Wednesday meetup.com event in Amsterdam. There I demonstrated how to use a passphrase on a Trezor hardware wallet device and what happens when you enter the wrong passphrase on a device. The video is available on my website at empoweredlaw.com.

The pros of a key or seed plus passphrase include standardization (though of course not all implementations follow the standards), great security if you use a sufficiently long passphrase; and division of control, in that a single seed or key alone isn't enough to compromise your assets. Some devices, like the Ledger hardware wallets, even allow you to type the passphrase directly into the device, minimizing the risks of a keylogger obtaining your passphrase.

Cons include user error and insufficient user knowledge of how to create strong passphrases. Some of the most common user errors include misspelling or mistyping your passphrase at set-up, not realizing that capitalization matters, and the most common user error: forgetting the passphrase and/or not backing it up. Unlike backing up your wallet, there are no software-based reminders nagging you to complete your backup. Passphrases, like passwords, are very easy to forget, especially if you're not using them regularly. And again, if you forget the passphrase you lose access to the asset. Period. Most security experts say a robust passphrase should include 6 or more words, not commonly found together (not a passage from your favorite book or song), that you can easily remember.

There are lots of resources to find ways to create a good passphrase. I wrote a bit about this in my free article Bitcoin Security Made Easy: simple tips for non-experts [https://medium.com/@pamelawjd/bitcoin-security-made-easy-simple-tips-for-non-experts-10f1954634d4].

Key plus passphrase is an excellent solution for many cryptoasset owners. Many of the most popular hardware wallets, including Trezor, Ledger and KeepKey, have optional passphrase functionality built in (usually as an advanced feature). With minimal training, it's fairly easy for a non-technical person to implement and use a passphrase feature. From a security perspective, if you use a sufficiently long, unique passphrase, it would be almost impossible for someone to brute-force it. From a resiliency perspective, it's fairly easy to build resilience into the system, simply by making more copies of the key/seed and making more copies of the passphrase and storing them securely. From a recovery perspective, however, using a passphrase requires an additional step. If you use a passphrase you *must* back it up, otherwise your heirs will not be able to access your assets. The passphrase backup must be stored in a separate location from the seed it's related to.

You can use more than one passphrase on most passphrase-compatible wallets. Sometimes these additional passphrases are used for duress, emergency transactions, or even as a way to allocate assets among heirs. To learn more, see passphrases for duress [https://blog.trezor.io/hide-your-trezor-wallets-with-multiple-passphrases-f2e0834026eb].

Shamir's Secret Sharing

The third option is called Shamir's Secret Sharing, or SSS. Adi Shamir invented this scheme, and of the three technology-based methods mentioned here it is the most cryptographically secure. In simple terms, SSS allows you to take a secret you know (your seed) and divide it into N parts, of which M will be needed to reconstruct the secret in full. Sounds a bit like multisignature, right? Conceptually it is, except there's a very big difference. In SSS, one secret is divided into parts. In multisignature, several small secrets are combined together to make something that's not a secret: a public address. Multisignature can be combined in a way where one person never knows all of the components of the secret, SSS cannot.

One of the most interesting things about SSS is that knowing one part of the secret, mathematically, gives you no information about the rest of the secret. This makes it ideal as a backup tool for cryptoassets. Each share is independent and does not contain more or less information than any other share.

Other pros include flexibility, For example, you can set up a 2-of-3 scheme, where any two pieces can recreate your secret in full. In this way, SSS resembles multisig. However unlike multisig, there is no on-chain accountability. In SSS, the pieces are both split and reassembled using software, offline. Once you've successfully reassembled the seed, you would use it to sign a standard transaction and broadcast it to whatever network you're using. There would be no evidence that you used SSS on the blockchain; you wouldn't know which pieces came together to reconstruct the secret; it would look like any other transaction.

Why don't most wallets offer a SSS backup as an option today? Because there are a few big cons. The biggest con is that we are still in the very early days of cryptoasset technologies. SSS is an advanced feature; it's easy for newbies to make mistakes and unknowingly make their secrets unrecoverable. When I talk to wallet companies, I often hear, "It's hard enough to get people to make a single backup! Now you want them to go through a whole 'key splitting' process?" Using these features requires advanced knowledge, an understanding of why people should use them, and what to do once they have used them—far beyond the service an FAQ or customer service ticket desk technician can provide. It's not considered an essential security feature for most wallets, and in fact increases complexity. Another con is lack of industry standardization.

Because these features aren't offered in many wallets, some people implement SSS themselves using code they write or free software available online. Unless you are a skilled developer, able to read and understand code, do not use the Shamir's Secret Sharing software to divide your cryptoasset keys or seeds. **Never enter your cryptoasset keys or seeds into unknown, untrusted software.**

The Trezor hardware wallet company has proposed standardizing the implementation throughout the industry. You can find out more about the standard at SLIP0039 [https://github.com/satoshilabs/slips/blob/master/slip-0039.md] (Satoshi Labs Improvement Proposal) on GitHub.

Today, Shamir's Secret Sharing is an excellent solution for the most tech-savvy cryptoasset owners. Tomorrow, hopefully, it will be a viable option for everyone. In theory it's easy to use, but in practice it's a bit tricky to use correctly. It does provide the most cryptographically-secure separation of powers backup method, and it's fairly easy to build robust resiliency into this system, simply by making more shares of the key/seed or making more copies of each of the shares and storing them securely. From a recovery perspective, a high-quality, tested implementation of SSS is one of the most recoverable splitting options available.

Physically Dividing the Key

And now we come to the easiest and the absolute worst option: physically dividing keys or seeds by splitting them yourself. Most people literally cut a paper in half or quarters and then give pieces to their loved ones. It's so easy to do and, if you don't actually understand cryptography or the security model of cryptoassets, it makes perfect sense. But let's explore why this is actually the worst option, from both a security and resilience perspective.

Your wallet was designed with security and resilience in mind. If you think about it, your wallet has only three jobs: allow you to receive assets, secure them, and send them to others. A wallet company that fails at any of these three things won't be around for long, and so they take these things seriously. When they choose a security strategy, they make choices that they believe will protect you best while using their product. When you divide a key on your own, it's like you're taking security into your own hands; you're independently redesigning the security model, which is never a good idea, even if you're a professional.

When people physically divide their keys, they usually don't note the software or total number of pieces, on each piece. They're going for "security by obscurity", with logic that goes something like, "If they don't know what this is, then they won't be able to steal my cryptoassets." Most people feel particularly clever when they do this too, like they've really protected themselves well. Unfortunately, that is not the case. By using non-standards-based key division techniques (that's a nice way to say cutting up your keys), you've actually increased risk of loss.

Cutting your keys or seeds in half reduces security, by a lot. If you cut your 12 word seed in half and give me only one half (6 words) you may think your seed is now half as secure, meaning that it takes 50% of the time to find 6 words that it takes to find 12 words. In fact, it is approximately 18,446,744,073,709,551,616 times easier to find 6 words than it is to find 12. Of course, the exact number depends on which half you gave me, specifically if it involved the checksum or not. From a technical perspective, this level of security is considered crackable within the range of today's computers. You've changed the security level from "not even in a million years" to "possible given enough hardware." While the people holding these pieces might not be savvy enough to do this on their own, why put them in a position to put your assets at risk?

On the flip side, if you have your backups divided in non-standard ways, it makes it more difficult for honest helpers to actually help your heirs recover your assets. There are so many ways to divide keys; without an understanding of what you've done, your heirs and helpers could be on a multi-year quest to solve the riddle of the divided keys. They might just give up.

Dividing your keys yourself is a terrible solution for most cryptoasset owners. While it's arguably the easiest to execute, from a security perspective, it's insecure. From a recovery perspective, it's a disaster. While it's your money, and you can store your backups any way you like, it seems reckless to choose this option when the others solve the same problem in a more secure *and* more resilient way.

New Division Options Are Coming

Bitcoin's advanced scripting and Ethereum's smart contracts can both be used to separate powers in new and different ways. But just because a feature exists in the protocol doesn't mean that you can see it in an easy-to-use user interface or application. In fact, most of these features have not been developed into user-friendly applications yet. For more on this, see Smart Contracts For Inheritance Planning.

Don't Build Your Own Encryption

No matter what division method you choose, or even if you choose none at all, if you're not a security expert you'll probably feel the temptation to encode your keys or seeds in some way. The thinking goes something like this: "If someone finds my seed words, they'll steal my assets, so I won't write them in order and put them in a safe—instead I'll let my creativity shine! I'll create my own secret code to protect my assets."

Then you might concoct an extravagant encoding scheme based on changing the words in some way—for example, changing the word "apple" to "sauce" "pie" or "jelly"—and then spreading the words out in a bunch of different places—say one word written in each of your favorite books on page 32. As you might imagine, these sort of schemes are disastrous when it comes to inheritance planning. But they're also terrible for your own ability to recover. This scheme, for example, relies on your memory to be perfect to remember the associations, long-term; for you to remember to say no when someone asks to borrow one of your favorite books; for your list of favorite books never to change; for you not to have a fire or flood; and so on and so forth.

One memorable example in the industry is when the organization We Use Coins lost approximately 7,000 bitcoin because they chose to create their own high-security scheme and ended up losing access to their private keys. As of April 2018, all of those coins are still sitting in the same address. You can read all about it here [https://bitzuma.com/posts/getting-the-most-from-your-first-bitcoin-purchase/].

These sorts of schemes are far more common than you might imagine. To date, I've seen seed words in poetry, music lyrics, books with words circled in them or written in margin notes, in journals (usually strewn about many different pages with no order), in drawings, and many other creative places. Creativity is awesome. It's just not awesome for this. Avoid getting creative with your backups, and save yourself and your heirs a lot of hassle, headaches, and loss.

Even some of my closest friends have done things like this, like my friend Jackie. Jackie knew me before I got into crypto and when we reconnected she asked me about my work. I told her all about bitcoin and how much I love working in the industry. She quickly became interested, and unbeknownst to me she decided to buy a bit of bitcoin. When she mentioned that she had bought some, I immediately asked about her backups (of course). That's when she proudly told me that she wrote her words down as part of a short story, where the third word in each sentence was a seed word. She hadn't told me or any of her heirs about this special story, so none of us would have been able to access her bitcoin if something had happened to her.

The point of this story is to let you know that lots of people do things like this. If that's you, don't be ashamed. Just plan to make a change as soon as possible. If you've given out pieces of your seed or key, consider securely creating new keys or seeds and sweeping the assets to the new addresses. Keep SURE in mind as you make decisions on when, where, and how to sweep those assets. If you've created a fancy encryption scheme of your own, destroy it. Create standard backups, like your wallet instructs you to do. You can use the backup templates in the appendix as a guide Keys, Seeds, and Access Code Templates. Oh and don't worry about Jackie, she's a backup maven now, just like you will be.

Choosing Your Separation of Powers Strategy

With that in mind, it's time to evaluate how many people or pieces should have to come together to reconstruct or recover your cryptoassets, and we need to balance that with the resiliency of your plan. Remember, this is *your* plan; it needs to serve *you*, with the knowledge you have today. It needs to honestly reflect your situation and those of your loved ones.

You might read the previous section and think, "That's all great but I'm not ready to do any of it" or, "I don't know enough to decide which strategy is best." In that case it's better to keep your backups intact, without dividing control, than to do it wrong and lose access completely. Do not implement something that you don't understand. If you decide to keep your backups intact, without requiring any other factor or piece, then you'll need very secure storage locations.

If you've decided you're ready to implement one of the division strategies then you need to make some choices. For most people, two separate pieces along with a couple of helpers is enough to provide great security and resiliency. If you've got 5 children and want to leave one piece with each, you could create a 3-of-5 plan (unless you think it's likely 3 will collude against the other 2). Unfortunately, this part of the planning can't really be done by the book. I can't tell you what's right for your situation. Because the right thing for you is completely dependent upon your unique circumstances. I spend a lot of time with my clients talking about these issues. I recommend you spend some time thinking about this, and later, in the SURE evaluation section, you'll revisit the idea of risk. It's also important to remember that you're not committing to this strategy forever. Your strategies can evolve as new technologies become more widely usable, as your knowledge and comfort with these technologies increases, and as your circumstances and relationships evolve. Today you're simply building something that you think will work for you and your family today. You're free to make changes any time.

Once you've finished consolidating your assets and you're happy with what you've got (or you're simply ready to move on) it's time to verify your backups and access codes, and then update your inventory and your Get It Done plan.

Audit Your Backups

The subject of security and cryptoassets is complex, nuanced, and highly dependent upon things like your individual tech know-how and comfort level, your ability to understand and evaluate the security risks you face, and how well you can balance the risks with the practicality of being able to access your assets when and how you want to. If you want to learn more about security (or anything else bitcoin related), start with Jameson Lopp's outstanding resource page:

https://lopp.net/bitcoin.html.

My goal in this book is to help you ensure that your assets can be accessed by your heirs when the time comes. For that, we simply need to make sure that you have backed up your access information and have stored it securely. We've already covered how to make backups, but if you need a refresher, see the Backing Up Your Wallet section of this book.

At minimum, you should audit all of your backups now. In this case, audit means to look carefully at every wallet you use. Make sure that you have properly backed up every wallet. Separately, make sure you have backed up any password, passphrase, or access codes needed to gain access to this wallet. Write the name of the wallet you're using on both the key or seed backup and any access code. If you're using a USB or electronic media backup, be sure to identify what is on the USB somewhere, perhaps on a label. Be sure you have at least two complete backups for every wallet and make sure they are stored in separate, secure locations; we'll discuss those next.

But before you move on, be sure you've updated your inventory sheet to reflect your latest changes.

Better Long-Term Storage

We've already discussed the importance of choosing access-controlled, fireproof, waterproof storage locations for both your key materials and your plans in the Get It Done: Making Your First Plan section of the book.

For most people, three to four secure storage locations are enough. Remember not to store your basic access plan or your legal documents, like your will, with a third party who will require your heirs to use legal process to access them, like in a bank safe deposit box. Store these items where your heirs will find them quickly when they need them.

Do not give your keys, seeds, or access code information to your lawyer to secure for you or to include in your will. For more information, see Make it Legal.

Storing Key Material

Now let's talk about how to store sensitive material at these locations. Most of my clients store their key materials in opaque numbered bags. These bags make it easy to tell if someone has accessed your secure storage space and compromised your key, and it also makes it easy to inventory. The bags are called cash bags or evidence bags, and they're available through Amazon, among other places. If you can't get opaque bags then you can put your items in a manila envelope and put that in the evidence bag. If you really want to get all James Bond, you can wrap your key material in aluminum foil, then wrap it in cardboard, staple that closed, then put it in an opaque, numbered, sealed envelope. But I don't actually expect you to do all that.

If you're storing USB or electronic media, you might want to consider buying anti-static bags. They're also available for sale on Amazon.

Access Logs

For each storage location you should have an access log. Each time you access the secure location, you'll log the date and time, and initial the log. If you need to access key material, you'll note that on the log too. By keeping an access log you will have a complete chain of custody, which will help you keep track of your key materials in an organized way. Here's a sample of what an access log might look like:

Table 3. Sample Access Log

	Date	Item Descriptio n	Purpose / Change	Bag#	IN or OUT	Signature
1	3/3/2017	Samourai Seed	Seed Storage	A47636283 4	IN	PM
2	9/17/2017	Samourai Seed	Recovery	A47636283 4	OUT	PM
3	9/17/2017	Samourai Seed	Seed Storage	A47636287 5	IN	PM

Location Chart

Depending on your situation, you might want to keep a separate storage location chart and leave locations out of your access plan. If you believe you face additional risks from someone else seeing your plan and trying to use it to gain access to your assets, keeping a separate location sheet is a reasonable way to segregate information; just remember your heirs will need access to this information too, so store it wisely.

Update Your Get It Done Plan

If you've decided to update or change wallets, assets, or your storage locations, you'll need to update all of the copies of your Get It Done Plan. If you plan to finish the next section, Better Helpers, feel free to wait until you've finished it to do the update.

Better Helpers

In Get It Done, you identified some people or organizations that could help your loved ones identify your cryptoassets, access them, and help get them securely transferred to their rightful new owners. Now, we'll look more carefully at these helpers, identify their role in your plans, and try to add more helpers to the list.

In a perfect world, your heirs would be cryptoasset experts. They wouldn't need someone to help them with the technical or legal side of transferring these assets. They would know that using someone else's credentials to access their accounts could be a criminal offense in some jurisdictions.

A perfect helper is trustworthy, highly technical and into cryptoassets, great at explaining complex concepts, isn't afraid to ask questions, can self-educate and learn quickly, and has no interest in stealing your assets. As you might imagine, perfect helpers are hard to find. Even if you did have one, you always want at least two helpers, just in case. Two people can watch one another. Ideally they'll be people that don't already know one another, because that makes it less likely they'll collude. Two helpers are also better than one in case you misjudged a helper's technical or cryptoasset knowledge.

But what if you don't have a perfect helper. That's okay; with a combination of a few different people, you can sort-of build your own. Think of the people you know and trust. People who might be good candidates to be helpers. Even if they're not into cryptoassets, they might make a good helper. There are two roles that a helper can fill: trust and tech.

Begin by considering the helpers you identified in your Get It Done plan, if they're still people you trust. Even though they seemed like the best candidates at the time, you should evaluate them in the same way we'll evaluate new potential helpers. Make a list of potential trust and tech helpers.

What makes a good trust helper? Obviously they are people you trust, people who are trustworthy and honest. They're people who make good decisions, with strong values, and who are unlikely to develop an addiction to drugs, gambling, etc. They're people you would trust to watch your children, if you have them. People to whom you could give your house keys and not worry that when you came back your house would be destroyed and/or your valuables would be gone. Bonus if they wouldn't rummage around in your drawers and cabinets.

They're also people who aren't afraid of technology, are smart and curious, or very fast learners. Your trust helpers don't need to have an understanding of cryptocurrencies or cryptoassets, though it's great if they do. It's important that your trust helpers are able to ask your tech helpers the right questions and are not afraid to question authority, as a way to provide oversight. If they're fast learners and/or tech-savvy, they should be able to do this with minimal self-education. If none of the people you've identified has these characteristics, that's okay; just move forward to the next section.

Usually it's best if your helpers are not also your heirs. Selecting an heir to be your helper puts them in a position of power over your other heirs, which may or may not be okay for your situation. Think carefully about the power dynamics at play when selecting heirs as helpers.

Evaluating Potential Helpers

Only name people who are trustworthy. Then to decide amongst them, put an X in each category where the person has the quality, score 1 point for each X and total the scores for each person.

Table 4. Trust Helper Evaluation

Name	Tech-Savvy	Fast Learner	Questions Authority	Owns Cryptoassets	Total
Alice	X	X	X	X	4
Bob		X	X		2
Charles	X	X	X		3
Daria	X	X	X		3

Select two or three people from the list to be trust helpers. In our example, Alice looks like a great candidate, followed by Charles and Daria.

Now it's time to build a list of possible tech helpers. What makes a good tech helper? For this role, you'll need someone who is cryptoasset-savvy, someone who keeps up with what is going on in the industry, and who has owned at least two different cryptoassets. For example, if you've been holding BTC since 2013, then those keys also entitle your heirs to equivalent claims of BCH, Bitcoin Gold, and other associated airdrops or coindrops. You need a tech helper who is aware of new coins and assets, understands how to claim them, and understands which of your holdings might lead to other assets. Usually it takes years to develop expertise, so look for someone with at least two years of significant experience in the industry. Your tech helpers will most likely be the people who will teach your heirs how to hold these assets securely, if your heirs decide to keep them, and oversee the actual transfers, so it helps if they're good teachers too.

Many of you will have built relationships within your local community, and that means your tech helpers are likely to also come from that community. That's great but remember that people who know each other are more likely to collude, as they have built trust amongst themselves. This is why, ideally, you'll want someone tech-savvy in the trust role and/or tech helpers who do not know the other helpers — you want them to be independent. Remember, the goal here is to distribute trust and provide oversight, to make sure that no one person can steal from your heirs.

For your tech helpers, only name people who have years of experience owning more than two different cryptoassets. Then, to decide amongst them, put an X for each quality a person has, score 1 point for each X, and total the scores for each person.

Table 5. Tech Helper Evaluation

Name	2+ years	Good Teacher	Trustworthy	Independent	Total
Alice	X	?	X		2
Eva	?	X	?	X	2
Franklin	X	X	X		3
Gigi	X	X	X	X	4

From our list it looks like Gigi is an ideal tech helper, followed by Franklin.

We've now been able to identify Gigi, Alice, Franklin, Charles and Daria as helper candidates. If Gigi and Alice don't know one another, then we're in a very good position. If they do, we may need to consider naming another person or two to oversee the process.

Fiduciary Helpers

What if you don't have enough suitable helpers? Then you'll need a fiduciary to fill the vacancies. The term fiduciary, as used in this book, refers to a professional, usually a lawyer or accountant, who provides services to you as a client.

The word fiduciary comes from the Latin *fiducia* meaning trust. More than simply a trusted person, a fiduciary is someone who has a *legal duty to act in the interest of or for the benefit of another person*. The legal duty comes with oversight (e.g. from the bar association, the probate court, or a regulator) and the *highest standards of care*. In many jurisdictions, a fiduciary is prohibited from profiting from their position as a fiduciary. Very importantly, a fiduciary must prevent conflicts of interest, either with their own interest or between the interests of parties they represent. Finally, a fiduciary also probably has malpractice insurance.

If you can find a fiduciary who is qualified and willing to be your tech helper then your heirs are better protected, because if the helper does something to harm your estate (for example, steals the assets) your heirs can sue them and their insurance company. And as any good lawyer will tell you, two pockets are better than one.

The executor of your estate has a fiduciary duty to the estate and to the heirs to protect the assets within the estate. They will probably need your tech helpers to educate them, and the court (if a court is involved), about these assets when the time comes. Take a minute and think about who might fill this role.

Now that you've identified better helpers, it's time to decide if you want to contact them beforehand or not. The benefits of letting them know they've been selected include that you'll know right away if they're not interested in helping, and if they do want to help you have the opportunity to educate them about your plans and assets. The downside of letting them know is that if they are not trustworthy, they could use the opportunity to learn about your holdings and plan to steal them. You will need to consider how much information you share with your helpers, and when.

Remember to include enough information about your helpers so your heirs will be able to identify them. Names and email addresses aren't enough, use photos and links to online profiles to ensure they find the right person.

Revise Your Access Plan

At this point your Get It Done plan should need to be updated. Hopefully you've made some changes to your previous plan in terms of assets, storage places, or people. Be sure to destroy all of the old copies. By destroy, I mean destroy, not just shred. Secure document destruction can include burning completely, soaking in bleach water, or many other methods you can easily find online. Ideally, you'll complete the next section before you stop for the day. But if you need to take a break, then update your Get It Done plan and add a calendar event to remind you to come back and assess your plan.

Now that you've learned what makes a good plan, it's time to evaluate yours.

Evaluating Your Access Plan

When you set your goals at the beginning of the Make it Better section, you wrote down what your intuition tells you is your ideal balance for the four SURE elements. Now it's time to quickly assess your current plan. In the following sections, we'll do a detailed analysis of your current plan, but for now we just want to identify any glaring discrepancies. Without too much analysis, think about how you store your keys, where you store your keys, who knows about them, and the plan you created in the Get It Done chapter of this book. Now give your current plan a score from 1 to 10 in each section with 1 being Uh-Oh and 10 being Nailed It!

Table 6. Identify Your Current SURE Balance

	Security	Usability	Resiliency	Efficiency
My Goals				
Quick Analysis				

Detailed Analysis

For our detailed analysis, we're going to look at our current plan across three different dimensions: key material, places or storage locations, and people. For each of those dimensions, we will evaluate the SURE balance of our current plan and compare it to our goal. If there is a significant discrepancy, we'll need to make some adjustments. When all three dimensions of your plan are well-balanced and meet your goals, you have completed the task of Making It Better.

Keys, Seeds, and Access Codes

As part of your inventory, you should have a clear understanding of all of the keys, seeds, and access codes that you currently have in secure storage. We'll now evaluate how well-balanced your plan is with respect to key materials and SURE. Use the table below to record your scores for the evaluations that follow. Blank tables are also available in Appendix SURE Evaluation Template.

Table 7. Identify Your Keys & Codes SURE Balance

	Security	Usability	Resiliency	Efficiency
Keys & Codes				

Security of Key Materials

Consider the consequences of each of your keys, seeds, and access codes being compromised. Ask yourself, "If an attacker had this piece, what could they do?" Your goal is to have no single point of failure, meaning no one piece of information is sufficient to take any of your assets.

Next, consider whether possession of one of your keys, seeds, or access codes gives an attacker a way to find or compromise another piece of key material. Ask yourself, "With this, could they access my email account? Could they do a password recovery? Could they impersonate me?" Repeat the exercise for every software application, custodial service (like exchanges), or device that relates to cryptoassets.

Based on this analysis, give your current plan a score between 1 and 10 for the S component in SURE.

Usability of Key Materials

In terms of usability, for all of the following questions, fewer is better. How many different keys, seeds, and access codes do you have? How many different types of keys, seeds, and access codes do you have? (e.g. BIP39 seed, BIP38 encrypted key, keystore file, passphrases, passwords). How many different devices or codes are needed in order to get access to your assets? How many different software applications are you using? How many different types of tokens, assets, currencies do you currently own?

Based on this analysis, give your current plan a score between 1 and 10 for the U component in SURE.

Resilience of Key Materials

For each copy of each piece of key material, seed, or access code, consider the impact of the loss of that copy. If you lost that piece, would you lose access to the asset? If so, you'll want to increase the resilience of your key material and access codes. If not, congratulations — you've achieved resiliency against single point of failure. Now continue to evaluate if you have resiliency against two points of failure; that is, are there two pieces whose loss would cause you to lose access to an asset?

Based on this analysis, give your current plan a score between 1 and 10 for the R component in SURE.

Efficiency of Key Materials

Look at your collection of key material, seeds, and access codes. Could you achieve the same or better levels of security, usability, and resilience with fewer items? Could you consolidate your assets? Could you reduce the complexity in terms of distinct pieces of information of storage, without negatively affecting the other three elements of SURE? In other words, is this the simplest arrangement of key material that achieves your goals?

Based on this analysis, give your current plan a score between 1 and 10 for the E component in SURE.

Identifying Areas of Improvement

Before we move to the next section, take a few minutes to think about all four components of SURE as they relate to your keys, seed, and access codes. Identify where there are big gaps between what you wanted to achieve and where you're currently at. Write down a few things you want to do to improve your SURE score for your keys, seeds, and access codes.

Evaluating Your Storage Locations

Now we'll move to evaluating your current storage locations, the secure places you've chosen to store your key material, seeds, and access codes. You can use the table below to record your scores for the evaluations that follow.

Table 8. Identify Your Places SURE Balance

	Security	Usability	Resiliency	Efficiency
Places				

Security of Places

When evaluating the physical security of each location, there are three primary considerations: (1) difficulty of access (what are the barriers to access?), (2) likelihood of exposure (will anyone know they've gained access?), and (3) the severity of punishment for unauthorized access of this location. You'll need to evaluate these three things from the perspective of both an outside attacker and an insider or someone who knows you.

For each storage location, ask yourself the following questions: How difficult is it for an outside attacker to gain access to this location? If they did gain access, would I know? For example, is there video surveillance at this location that I review regularly? If I did notice, what is the punishment? For example, robbing a bank is punished more severely than burglarizing a home. Now ask yourself the same questions from the perspective of an insider, someone who knows you.

How likely is it that the security of this location will be breached by a natural disaster, like an earthquake or volcano eruption, fire, flood or water? How likely is it that you will face political or legal threats to your assets at this location? Many people who face these sorts of threats pick secure storage locations at a great distance, even abroad.

Based on this analysis, give *each location* in your current plan a score between 1 and 10 for the S component in SURE, and based on those scores, give your current plan an overall score, again between 1 and 10, for location security.

Usability of Places

Assuming the time has come for your heirs to take action, how difficult will it be for them to find and access your plan and each storage location? Will they be able to locate your key material, seeds, and access codes easily once they have access to the location? For example, if you've buried a Cryptosteel in your backyard, will they spend years digging? Can your heirs afford to travel to the storage location? Do they have the requisite identification, perhaps a passport or a government ID, to access the location? Will they need a court-issued document in order to gain access? And if so, how long does that usually take to get?

Based on this analysis, give *each location* a score between 1 and 10 for the U component in SURE, and based on those scores, give your current plan an overall score, again between 1 and 10, for location usability.

Resilience of Places

Perhaps the area you live in is prone to natural disasters of such breadth that you need to consider distant storage sites. Take each location individually and imagine it was demolished by a natural disaster, the kind that takes out a large geographic area. Would something like that also take out any of your other locations? How many would be affected? Would your access plan be able to survive a local disaster? A regional one? A national one?

Based on this analysis, considering all of your storage locations together, give your current plan a score between 1 and 10 for the R component in SURE.

Efficiency of Places

What is the annual cost of maintaining your storage locations? Be sure to include the cost of traveling to each location, at least annually, to complete an annual audit of storage materials. Are there maintenance costs, such as rental fees, that could be lowered? Could you achieve the same levels of Security, Usability, and Resilience with a simpler, less expensive system? For example, vault and safe deposit boxes often come in different sizes. Much of your key material, seeds, and access codes will be small. You can save money by selecting a smaller sized box. Another example might be to reduce the number of storage locations, say from 7 to 5, but choose more diverse locations.

Based on this analysis, considering all of your storage locations together, give your current plan a score between 1 and 10 for the E component in SURE.

Identifying Areas of Improvement

Before we move to the next section, take a few minutes to think about all four components of SURE as they relate to your storage locations. Identify where there are big gaps between what you wanted to achieve and where you're currently at. Write down a few things you want to do to improve your SURE score for your storage locations.

Evaluating the people

Now it's time to consider the people involved in your life, your plan, and those who might find you a nice target. Use the table below to record your scores for the evaluations that follow.

Table 9. Identify Your People SURE Balance

	Security	Usability	Resiliency	Efficiency
People				

Security of People

How secure is your plan when it comes to those who live in your home? If someone you live with developed a gambling or drug problem and needed money to feed their addiction, would your cryptoassets be safe? Imagine you have a large pile of cash in your home safe; do you feel confident that the cash would still be there if those people were alone in your home for a week? What if they knew the location of your home safe?

Have you been talking about your cryptoassets on Facebook? Do your neighbors and acquaintances think you are "cryptorich?" How secure is your plan when it comes to third parties that might target you? What about third-party opportunists: could someone casually pick up one or more of your keys, seeds, or access codes?

Who has access to each of your locations? What would happen, for example, if an employee at your lawyer's or accountant's office came across your file? Would they have enough information to steal your assets?

Are the people you've chosen to share your plan with also your heirs? If so, are you vesting too much power in one person, or just a couple of people? If you've chosen two people, what are the chances they collude to either hold the funds for ransom or steal them from your heirs? Could your heirs seek help from outside third parties, like courts, if this happened?

Based on this analysis, give your current plan a score between 1 and 10 for the S component in SURE.

Usability for the People

While for inheritance planning purposes the more knowledge your heirs have the better, in order for a more peaceful, happy life while you're here, sharing everything might not be right for your specific situation. If your heirs do not share a similar tolerance for risk, they may not easily accept the volatility of cryptoassets. If you believe they would want to exert influence over your holdings, you might prefer to tell them less, or nothing at all. That said, most people choose to share some information with at least one other person during their lifetime. Could the person or people you've chosen to share your plan with understand and use it themselves? Would they need help from others? How much help would they need? Are they interested in learning more about these assets? Do they have the technical acumen to oversee helpers? Could you help them develop the skills needed to effectively oversee helpers?

If you've decided to train someone to use your plan, first go over the plan with them, answering questions as you do. Then, wait a couple of days before having them pretend to execute the plan as if you weren't there. This exercise will help you both identify knowledge and instruction gaps, and help them build confidence.

Based on this analysis, give your current plan a score between 1 and 10 for the U component in SURE.

Resilience of People

Often the people we choose to share pieces of our key materials and plans with are the closest to us, our spouses, children, best friends. They're the people we vacation with, those we celebrate holidays with. Could your plan survive if one of these people were to pass before you or at the same time as you? Are the people so close to you that if something happens to you it's likely to happen to them too? Do you want your plan to survive them?

Based on this analysis, give your current plan a score between 1 and 10 for the R component in SURE.

Efficiency of People

Now it's time to look at efficiency when it comes to the people involved in your plan. Could you achieve the same result with fewer people? Could they achieve the same result with less knowledge? Could your heirs achieve the same effect with less help from outsiders, without sacrificing the other levels of SURE?

Based on this analysis, give your current plan a score between 1 and 10 for the E component in SURE.

Identifying Areas of Improvement

Now it's time to take a few minutes to think about all four components of SURE as they relate to the people involved in your plan. Identify where there are big gaps between what you wanted to achieve and where you're currently at. Write down a few things you want to do to improve your SURE score for the people who will be using your plan.

Improve Your Access Plan

Based upon your SURE analysis of keys & codes, places, and people, update your overall evaluation of your current plan in comparison to your goals.

Table 10. Update Your Current SURE Balance

	Security	Usability	Resiliency	Efficiency
My Goals				
Current Plan				

Based upon your overall SURE analysis, it's time to make some decisions. What changes do you need to make to your plan in terms of security? Decide what are the most important security upgrades you need to make in terms of keys, people, and places, and set aside time on your calendar to make them. If you need to do research about a specific wallet or separation of control method, block out the time you need to do it now. Consider the changes you want to make to improve usability, resiliency, and efficiency and create a plan and timetable to get them done.

Finally, review the actual access plan you have in writing and update it today. Start with the letter you're leaving for your loved ones: is it accurate? If not, update it. If you've decided to talk to your loved ones or your helpers about your plans and your cryptoassets, be sure to update them too. Don't forget to update supplemental information, like storage location logs or password managers. Continue to make changes and update your plans until you're SURE they match your goals.

Consider setting some time aside to educate your heirs about technology. Whether or not you decide to talk to them about your plans, you can teach them some basic skills, like how to use a password manager or 2FA, and talk to them about encrypted communications. Build your heirs tech skills as a way to prepare them to take on cryptoassets if they need to.

By now your plan should have evolved to the point where you are reasonably sure your loved ones will be able to access your cryptoassets with the guidance you've provided.

Additional Legal Documentation Required

A great access plan alone isn't enough to ensure your cryptoassets are passed to the people, charities or other beneficiaries you choose. For that you'll need a legal will or trust, conforming with the local laws where you live. In the next section, Making it Legal, you'll learn about wills, the laws you need to know about, how to find, choose, and fire a lawyer, how to save money on legal fees, and much more.

Make it Legal

In the previous sections, we created an access plan so that your loved ones could inherit your cryptoassets. But who exactly will inherit them depends on you—and what you do or don't do. You have the power to decide who will get to inherit your assets and who won't. But it requires action.

The Laws You Need to Know About

Cryptoassets will become part of your legal estate in most jurisdictions, and will be subject to local laws and taxes, whether or not you want them to be. This is the reality of our current jurisdiction-dependent judicial system. Your access plan alone won't be enough to ensure your assets are passed to your chosen beneficiaries. Without a legal plan, the default laws of your jurisdiction will usually apply. In the next section, we'll look at these default laws and ways to ensure *you* get to decide who gets what. We'll also talk about taxes, what really happens in estates, and how to avoid the most common mistakes people make in estate planning.

Consider Who Gets What

Before we get started, however, take a few minutes to think about who you would like to inherit something from you—who will be the beneficiaries of your legacy. Right now we're not going to think about what or how much each person should get, the point is to make a list of the people you'd like to get *something*. These questions go beyond cryptoassets — they are applicable to all of your assets.

Begin by considering the people closest to you. Do you want them to receive something? If so, add their names to the list. Do you have a responsibility to support anyone? Are there people in your life that could really benefit from inheriting your cryptoassets? What about your mentors, best friends, colleagues, or neighbors? Are there charities or other organizations that you'd like to give something to? Don't worry that your list is too long or too short; for now it is what it is.

Now look at your list. Are there people who are not your spouse, child, parent, or sibling there? If so, you need a will, testament, or trust. Let's talk about why.

Intestacy Laws and Why You Should Care About Them

Did you know that if you die without a will, or testament as it's sometimes called, local law decides how your assets are divided?

In the USA, and often abroad too, these laws are called laws of intestacy, and most of them only recognize formal legal relations — meaning relations by blood or marriage. Your long-term, live-in domestic partner? No. Your best friend since preschool? No. Your children through marriage? No. Your favorite charity, community center, or political cause? No.

Even if you're not concerned about how your assets will be allocated, if you have children, you should seriously consider writing a will to designate someone to care for your children if both parents die. While that has nothing to do with cryptoassets, it's something those of you with children or dependents of any kind should think about.

These laws often result in unexpected and undesirable outcomes. Let's look at an all too common example. Taylor and Pat are married, with three children and a dog. Well, actually their first child Lisa is not Taylor's biological offspring. Taylor and Pat met shortly after Lisa was born. They fell in love and have been together ever since. The family never really discusses that Lisa isn't biologically Taylor's; no one seems to care. They think of theirs as a normal family, nothing extraordinary; Taylor and Pat never get around to writing a will. Unfortunately Pat dies, and the entire family is grief-stricken. Legally, in their jurisdiction, everything goes to the surviving spouse, in this case Taylor. Taylor inherits everything and from a practical logistical perspective, nothing much changes. Taylor still cares for all three children and the dog.

Taylor lives for five years and never remarries. Then Taylor dies. What happens to the family assets? Well, in most jurisdictions, Taylor's two biological children inherit everything and Lisa gets nothing. Why? Because the laws of intestacy only recognize legal and blood relationships. Unless Taylor went through the formal legal process of adoption, through the state, Lisa is not technically related to Taylor. Imagine the problems this could cause; the anger, the resentment. The siblings could be fighting forever. Even if the siblings all agree that Lisa should get one-third, and they share their inheritance with her, there could be costly tax implications because Lisa's share isn't technically an inheritance.

Then there's the issue of the family dog. Who should care for it? Will someone be compensated to care for it? This might seem like a small issue, but it could turn into a war. Sam Simon, the wealthy philanthropist co-creator of the Simpsons, left his dog to friends when he died. But it seems he didn't allocate funds to care for the dog in the manner to which it was accustomed when Sam was alive. According to the news, those friends then sought $170,000 from Simon's trust as compensation for caring for the dog for one year. In order to collect the funds, the friends would need to sue the trust, leading to a messy and costly legal battle. Under intestacy laws, pets are considered property, not family. And you know what happens to unwanted property. Many pet owners are now providing for the financial and physical care of their pets in their estate documents, because intestacy laws leave them out.

Intestacy laws were written for another age, another time.

Don't rely on them to distribute your assets. Make a choice to decide for yourself where your assets will go.

What You Need to Know About Wills and Probate

The simplest way to avoid intestacy laws is to write a will. A will is simply a written document where you declare who should inherit your property. Sometimes it can even be handwritten. Just like there is a statute in your jurisdiction covering intestacy laws, there is one covering wills. The statute in your jurisdiction will say exactly what needs to be in the will—for example, you must sign and date it—in order to make it a legal document; the requirements vary from place to place. Handwritten wills are often called "holographic" wills, and they are recognized in some jurisdictions but not in others. Holographic wills and wills written without the assistance of a lawyer are often troublesome. Frequently, people fail to account for all of their assets, leave out important people, or don't put the required language into the will, leaving its terms ambiguous or the document unenforceable. Too often, errors, omissions or ambiguity in wills result in long legal disputes among heirs, and the money saved by doing it yourself is spent many times over fighting in probate court.

Probate is the court supervised process that ensures your assets get to your legal heirs. Depending on your jurisdiction and your assets, there may be both formal (long and expensive) and informal (fast and inexpensive) process options. The probate court has jurisdiction over the process and oversees the administration of your will; this is how your heirs become the legal owners of your assets. Let's look at an example of how the process works.

When a company goes out of business, there is something called a winding-up period. During this period, the owners or executives do all of the things that need to be done when a business closes. Creditors are paid, customers and suppliers are notified, physical property is sold or donated, bank accounts are closed and the business is officially closed. When a person dies, a similar process takes place, and someone, usually called a legal representative or executor, does all the work of notifying people, distributing property, closing bank accounts and paying creditors, on behalf of the deceased person. Legally this is an important point. Most assets do not automatically transfer to beneficiaries, and instead they must be transferred by way of some legal process. The probate court often oversees the process. Let's talk about how to transfer ownership of assets both before death and after.

How would this process work with cryptoassets? What if I want her to inherit all of my bitcoin? From a practical perspective, I could just give her copies of my keys. But that will result in her effectively owning the bitcoin now and that's not really what I want. I could make her promise not to use the keys, and even if she honored that promise, she would have to keep the keys secure — something that even avid cryptocurrency enthusiasts find difficult.

Giving her a copy of my bitcoin keys is like signing the title to my 1965 Corvette, without designating anyone as the new owner. With a car, anyone who finds the signed title could simply write in their own name, or someone else's, as the new owner, present it to the department of motor vehicles, and effectively become the new owner of my car. With a cryptoasset, anyone who has the keys can present a valid transaction to the network and transfer the value to themselves or someone else. (By the way, I don't actually own a 1965 Corvette but I thought the reference would provide you with a nice visual image.)

With a car, it might be possible, with a lot of time, money, and effort to reverse the transfer, or I might be able to sue the thief, because their name would be registered on the title. With most cryptoassets there is no reversal and few have formal ownership registration. In order to sue the thief I would have to find their identity through some other means.

Cryptocurrencies are usually considered bearer instruments, like cash. Ownership is tied to possession of a credential, like a key, therefore tracking down someone to sue is much more difficult and often impossible. For cryptoassets that are intended to be securities or do have traditional ownership registrations, a more formal re-titling process, through probate, will probably need to occur. The legal issues surrounding how these assets will probaby be handled by your local probate court are best answered by a lawyer in your jurisdiction. I say "probably" because the laws have not caught up with technology and for now it's impossible to say, with certainty, how any probate court will rule on them. A good lawyer will be able to make a strong legal argument and convince the court to uphold your wishes. We'll talk more about finding a good lawyer later. For now, remember you need to be very careful with keys and ownership credentials. This leads us to one of the most common mistakes people make when it comes to key security and wills.

Don't Put Keys or Passwords in Wills

Wills Become Public Records

Imagine your login credentials, your private keys, becoming part of the public record. Available for everyone in your county or town to see. How long do you think it would take until the assets were stolen? My guess is that if your credentials are in your will, the assets will be gone long before your heirs can access them.

In many jurisdictions, wills become public records. Upon your death, your will is submitted to the appropriate court in the jurisdiction of your legal residence. For example, if your official residence is in Chicago, your will is submitted to the Probate Division of the Circuit Court of Cook County. As soon as one of your heirs submits your will to the court, it will become part of the public record. This means anyone can obtain a copy of it so long as they follow the court's procedures. This is a huge security risk. Legally, your heirs will not be allowed to transfer these assets immediately, putting your cryptoassets in jeopardy, and increasing the likelihood of theft. Courts are slow, and methodical. They want to get it right, there are big incentives to get things right, and few incentives to move quickly. They tend to err on the side of accuracy — slow, slow accuracy. Few courts would recognize the urgent need to transfer cryptoassets immediately. Few heirs would either.

Avoiding a Race To The Jackpot

By including your key material or access credentials in your will itself, you might be creating a race to the jackpot. A race where the first person that finds your will and understands these cryptoassets might be tempted to transfer the assets immediately to their control. With minimal effort by the thief there would be almost no way to prove the assets weren't stolen by some random "hacker" and no-one would know it was an inside job.

Attorneys' Office Documents Aren't Secure

Additionally, if you are having a lawyer write your will, your keys could be compromised long before your death. Most attorneys use document management systems, which provide access to various people in the firm, including the IT department. These systems are fantastic for collaboration in the office, but in the case of cryptoasset credentials, collaborative software is the enemy of security.

Just think about how many people in the firm will have access to your keys. Do you trust them? All of them? Many of these systems do not use internal encryption, and they back up to cloud servers. Even if you do trust them, do you trust that they won't get hacked? Few attorneys have a good command of operational security; fewer have a good understanding of key management and cryptoassets. As a lawyer, I rarely make unequivocal statements, but this time I will: Do not give your access credentials to your lawyer to include in your will.

Period.

This doesn't mean your attorney can't securely hold an opaque, numbered, sealed, envelope that you provide to them that contains a passphrase or access credential. But do not give them the keys or credentials to type into their computer system or print out for you.

Felonious Heirs

Often laws aren't as clear as we hope they'd be. For example, the Computer Fraud and Abuse Act (CFAA 18 U.S.C. § 1030) has been interpreted in many different ways by many different courts throughout the USA. There is confusion among the courts (and the people) as to what actually constitutes "unauthorized use" of a computer, and many, many articles have been written about this confusion. You can read about some of these cases at the National Association of Criminal Defense Lawyers [https://www.nacdl.org/] association website and on the Electronic Frontier Foundation's [https://www.eff.org/] website.

So what does the CFAA have to do with cryptoasset inheritance planning? Almost every online service publishes terms and conditions (T&C) or terms of service agreements (TOSA) and most of these prohibit anyone but the account holder to access the account using the account holder's credentials. If you're trying to prevent people from trespassing into other people's accounts, terms like this might make sense. But for inheritance purposes, if you ask your heirs to use your credentials to access your accounts, you might unknowingly get them into legal trouble. While it seems unlikely that anyone would prosecute someone for accessing assets that they have lawfully inherited, consider for a moment what happens when one of the other heirs, who didn't get what they expected, decides to use an accusation of "unlawful access" as a tool in a legal battle over the assets.

This is one of the many reasons our plans provide for direct access to your cryptoassets, by using paper and electronic backups of seeds, keys, and access codes, rather than counting on your heirs to use your devices (and therefore your credentials) to access these assets. There's no reason to add another layer of complexity to the process; it's already complex enough.

Please be aware that laws in your jurisdiction may provide penalties for directly accessing the digital assets of someone who is deceased. Your attorney can help you understand and refine your access plan to comply with local laws.

If your assets are held by third parties, these parties often have a legal obligation to work with your estate executor to transfer your assets. For example, if you have a traditional bank account, the way your executor transfers funds is governed by specific laws within your jurisdiction. The executor is not legally allowed to quietly use your login credentials to transfer funds without notifying the bank that you are deceased, closing the account, and complying with the banks policies for transfer.

The executor, too, is bound by the CFAA and all of the laws and organizational policies that restrict access to other people's digital assets. In response to this conundrum, the Uniform Law Commission, a non-profit legal research and recommendations group, drafted the Revised Uniform Fiduciary Access to Digital Assets Act or RUFADAA. The provisions of RUFADAA have been adopted in a number of states and those statutes usually allow the executor of your estate to legally access your digital assets and compel third party custodians to cooperate with requests to access digital assets—this would presumably include cryptocurrency exchange accounts, but only if the exchange has custody of the cryptocurrency. Remember, if you're using a decentralized exchange, there is no third party to compel. The application of

RUFADAA is beyond the scope of this book but it's worth your time to learn more, if you're interested.

Taxes

No-one likes to pay taxes. People and organizations all over the world hire tax lawyers and CPAs to help them optimize their tax strategy. In case you don't know, that's rich people speak for "avoid any and all taxes" within the letter (but not always the spirit) of the law. These strategies are usually legal. These strategies are why Facebook and other companies locate in Ireland (for example); they're minimizing their tax burdens. And while you might not be ready to move to Ireland, there are things you can do to minimize your tax burden.

It might help to consider taxes in three different categories: (1) those you pay when you're alive, usually just called taxes; (2) those your estate pays when you die, usually called estate taxes; (3) those your heirs pay when they inherit value from you, usually called inheritance taxes. An excellent tax professional can, and should, help you with all three.

It's important to remember that people's goals differ. For some people, the most important thing is leaving something for their loved ones. For others, it's minimizing taxes they have to pay this year and next. Your specific situation will dictate your goals, and a good tax professional will help you identify your goals, explain options, and help you create a strategy to reach them. For example, in the USA many people today are using charitable remainder trusts to eliminate capital gains taxes while they're alive and give to a worthy charity when they pass. Of course this would mean that only the charity would "inherit" the assets in the trust, not any personal heirs but the tax benefits while you're alive are great. If you're charitably inclined and depending on your situation, this might be a good option for you. The point is not to suggest any one solution, but to encourage you to do some research and evaluate if working with a professional is right for you.

That said, there's a big difference between working within the law, for example, to pass assets to your heirs outside of probate, and passing assets to them outside the law. The reason not to do this isn't because you're a good citizen and you want your family to pay their taxes, although this statement may be true. The reason you probably shouldn't do this is because you could, unintentionally and unknowingly, be making felons of your heirs.

In many jurisdictions, you are required to declare inheritance proceeds, but not necessarily pay tax on them. In the US in 2018, the IRS allows [https://www.irs.gov/businesses/small-businesses-self-employed/frequently-asked-questions-on-estate-taxes] federal taxpayers to pass approximately $10 million to your heirs without your estate needing to pay tax. If your estate was worth $12 million at the time of your death, only the excess $2 million would be taxable. Some states in the USA have imposed their own inheritance and estate taxes but the most have not.

Secret Transfers

Inevitably, whenever I talk about inheritance of cryptoassets, someone says, "I'm just going to pass them to my heirs. No paperwork and no taxes." Many believe that these assets can easily circumvent the taxing authority in their jurisdictions. The idea is if you don't tell the authorities they won't know. In some jurisdictions and for some families this may be true, but for the vast majority it's not. Failure to pay taxes carries high penalties in most jurisdictions, including forfeiture and jail time. You won't be around to be penalized, but your heirs will. They will be the ones who have to hide, they will be the ones who have to figure out how not to get caught. Are your heirs comfortable doing that? Are they savvy enough? Are they *all* savvy enough, and are they likely to remain united in this forever?

Consider this situation. Imagine you're extremely wealthy.

Imagine you have $25 million in cryptoasset holdings. You have five children and each of them inherits $5 million in undisclosed assets. Do you think they'll go on a shopping spree? Can you picture the photos on Instagram or Facebook of their lavish travels, new house, and new car? Their inheritance probably won't be a secret for long. Worse yet, if one of them spends all of their money, they'll have the perfect leverage to get the others to give them part of their inheritance. They'll say something like, "I have nothing to lose, so either you give me some of your money or I'll turn you in for not paying taxes." In an effort to avoid taxes, you might actually be making a choice for your heirs that ends up being a curse instead of a gift. Think carefully about making choices that could affect your heirs in this way, and seriously consider talking to a lawyer about tax strategy, so that whatever choice you make is an educated one.

Legal Trusts

There are ways to keep your assets out of probate completely, for example by creating a legal trust. A trust is similar to a holding company or corporation. Legally it's considered an independent entity. Assets the trust owns would not be administered by the probate court, because they're not assets you personally owned. Creating and funding a trust is not usually a do-it-yourself type of thing. There are many different types of trusts, and each has its own quirks and specific requirements. Also, many trusts have their own tax reporting and payment obligations, so they may or may not be right for you. If you're considering a trust, do some research and talk to a knowledgeable lawyer.

Make Your Wishes Known

Remember, we are trying to prevent fighting amongst your heirs and to avoid legal problems whenever possible. If you don't make your wishes known, in writing, your heirs won't know what you want. That ambiguity leaves lots of room for misinterpretation, misunderstanding, and mischief.

I hope you now understand the benefits of having a written will. Whether or not you choose to make one is up to you. But if you do decide to move forward, it will probably be helpful to learn how to find and vet a good lawyer, and how to keep costs down; we'll discuss all of that in the next section.

Estate and inheritance planning are complex and nuanced legal areas. You wouldn't have your family doctor perform your heart surgery, you shouldn't hire a general practice lawyer to do your estate planning. You want a specialist, someone who is as passionate about inheritance and estate planning as you are about cryptocurrencies and asset tokens. It may seem impossible, but it isn't!

How To Find An Estate Planning Lawyer?

Unfortunately, inheritance laws are still based on physical jurisdictions, so you'll need to find someone who is licensed to practice law where you live. In the USA, estate laws are *state specific* and yes, there are different laws in every state, which is why it's important to get someone with knowledge of the quirks in your jurisdiction's laws. The best way to find a great lawyer is from personal referral, so ask your friends, family and acquaintances if they've got an attorney they love. If the recommended attorney isn't a specialist in estate planning, ask the attorney to refer you to a colleague who is. Many states require lawyers to be licensed. Often, but not always, the state's bar association administers and maintains attorney licensing.

Check Public Discipline Records

In most jurisdictions, lawyers must obtain a license from the state in order to practice law. These licenses require lawyers to not only follow the law, like any other citizen, but also to follow the rules of professional conduct, as determined by the highest court in the jurisdiction. The rules of professional conduct are long and complicated, and are intended to provide guidance to lawyers in the practice of law, and to protect client interests. The rules cover things like communication, confidentiality, and conflict of interest. If a lawyer violates the rules of professional conduct, they can be disciplined privately or publicly, and can face sanctions like suspension of their license to practice law or even disbarment, which means they'll never be able to practice law in that jurisdiction again.

Let's see how these rules can work for clients. Pretend that three months ago you hired an attorney to write a will for you. You signed a retainer agreement, to hire the lawyer, and paid them a retainer fee of $2,500. During the initial meeting your lawyer said they'd have a draft will for you to review in two weeks, but you never received it. In fact, you've heard nothing from them since that initial meeting, despite calling and emailing repeatedly. You've made four appointments to see the lawyer again, to find out what is going on, but the lawyer's office manager has cancelled the appointment at the last minute every time, giving one excuse or another. Now, even the office manager is ignoring your requests for an update. What can you do? If you're in Illinois, Rule of Professional Conduct 1.4(a)(3) & (4) [http://www.illinoiscourts.gov/SupremeCourt/Rules/Art_VIII/ArtVIII_NEW.htm] requires lawyers to keep their clients "reasonably informed about the status of the matter" and "promptly comply with reasonable requests for information." Based upon our facts, you would have every right to file a grievance (or complaint) with the licensing organization, for violation of the professional rules of conduct.

In Illinois, the organization is the Attorney Registration and Disciplinary Commission or ARDC. To find the organization responsible for licensing attorneys in your jurisdiction, try searching with your location and the words "attorney grievance." Using the privacy-focused DuckDuckGo search engine, the query "Illinois attorney grievance" returned the ARDC as the fourth result. If you visit the ARDC website you'll see how to file a grievance. In your grievance, you would explain that your lawyer has stopped communicating with you, and the bar association would initiate an independent investigation to see if the rules have been violated. Filing a grievance should only be done as a last resort; it's a very serious matter. Don't file a grievance if your lawyer doesn't return your call within an hour, and don't threaten to do so either. If you become known as a client who files frivolous grievances, lawyers may decide not to accept you as a client. Carefully read the instructions for filing a grievance, review the appropriateness of the matter, and only do so when it's absolutely necessary.

Lawyers who violate the rules may be publicly disciplined by the licensing body, and if they have been you can find those records online. Usually you'll search for a lawyer's record and find no disciplinary actions. If you do find a disciplinary report, read it and see if it's relevant to your issue. Not all lawyers who have had public discipline are bad lawyers and not all lawyers without public discipline are good lawyers. This is simply one more piece of information, one more tool, for you to evaluate your lawyer.

Interviewing Prospective Lawyers

Many people hire the first lawyer they find, probably because the process of finding a good lawyer isn't fun. But it's important that you try to find a few different lawyers and interview them. You are the client. You need someone who you feel comfortable with, someone who you can communicate with, someone you can trust. After all, you'll be sharing many of your private, personal details with your estate planning lawyer. Don't settle for the first lawyer you talk to: you deserve choices and options. Interviewing a few different lawyers will also help you learn how to keep costs down.

Research your lawyer before the interview. Read their website carefully, look for typos and inconsistencies. Research their disciplinary record and look for reviews online.

Interview Questions for Your Lawyer:

- ☐ What percentage of your practice is estate planning?
- ☐ How many wills and/or trusts did you draft last year?
- ☐ Do you also provide tax planning services? If not, do you regularly work with any tax attorneys who provide these services to your clients?
- ☐ From this meeting until all documents are properly signed, how long can I expect the entire process to take?
- ☐ Please provide an estimate of costs for something like this
- ☐ Please describe your typical client
- ☐ Please describe your ideal client
- ☐ Are any of your current or former clients willing to talk with me or provide a written referral?

At the interview you are trying to get a feel for three major things: (1) how well the two of you communicate, (2) how important you will be as a client, (3) how honest this lawyer will be about fees and processes.

If you do not understand the words your lawyer is using, do not pretend you do. It's very important that you understand everything that's being said, and you should not feel ashamed if your lawyer is using words you don't know. It's their job to educate you, not your job to interpret what they're saying. If you don't know the definition of a word, or how it relates to you, simply say, "excuse me, could you please define that word?" or "excuse me, I'm not quite sure what that word means, could you please explain?" Even if you do understand every word, it's a good idea to ask them to explain something, because how they respond to the interruption will give you clues as to the type of person they are and whether or not you will be able to communicate well with them.

You're interested in learning how they describe their typical and ideal clients because, again, it will give you insight as to their professionalism and values. You want a lawyer who recognizes your importance as a client. If the rest of their clients are billionaires and you are a thousandaire, your matters may not be as important. But they might be. Consider the language they use to describe their clients carefully; do you want to be described in that way? Could they be describing you? If so, this lawyer may be a good fit.

It is impossible to estimate accurately how much your legal work will cost because the complexity of the matter only becomes apparent when the work is actually being done. That said, you're looking for a lawyer who takes billing and costs seriously. Someone who is interested in maximizing value for their clients and has enough experience to provide you with a fee estimate range.

Do not downplay your atypical, complex, unique family situation. Many clients want their lawyer to think they're "normal" (whatever that means) and so they fail to mention important people, situations, or assets. Adding people and assets after documents are finalized requires much more work and therefore will literally cost you money and time. Don't worry about your attorney thinking your family is dysfunctional. Almost all families are dysfunctional, and experienced lawyers know that already. **Great lawyers don't care how odd your family is, they simply want to make sure your estate plan is complete and accomplishes your goals.**

Now, let's look at how lawyers bill for their work.

Understanding How Lawyers Bill For Time

When you interview your prospective attorney, be sure to ask about their fee structures and ways to keep costs down. You've probably heard the term "billable hour" in reference to how lawyers bill clients for their time. This is the most common fee structure, yet the name is a bit misleading. Lawyers bill by the minute, not by the hour. Billable hour rates are usually divided into a per-minute rate with a minimum initial block of minutes of six, ten, or fifteen minutes for each interaction. As a purely hypothetical example, let's say that during law school I spent a summer working for a mid-size law firm. As a law student, the firm set my billable rate at $125 per hour (or $2.08 per minute) with an initial 6 minute billable block. Each time I called a client and left a voicemail message, about a 2 minute endeavor, the client was billed for 6 minutes, or roughly $12.50 per voicemail. One way to keep costs down is to avoid having your attorney chase you for information by being responsive to their requests, and thereby reduce the number of interactions. Another strategy for reducing costs is to ask if a more junior attorney or paralegal can do some of the work on your matter. The hourly pay rate for staff is usually significantly lower than your attorney; lower hourly rates mean usually mean lower bills, unless the staff is much less efficient.

Not all lawyers bill by the hour for all matters. Increasingly, fixed fee agreements are becoming much more popular, as clients seek to really understand how much their legal services will cost before incurring them. Fixed fee agreements are exactly what they sound like: you pay a fixed fee for a fixed product. They're quite common for drafting simple documents, like a will or power of attorney. However, unless you have a very traditional family and a minimal number of traditional assets, your estate documents might not be so simple. While you might start out with a fixed-fee will, extras might add up quickly. You may have heard of contingency fees or "you pay only if we win" sort of fee structures. These are not applicable to estate planning; they're prohibited by the rules of professional conduct.

What is a Retainer Agreement?

Often at the end of the first meeting, the lawyer will ask the client to sign an *Engagement* or *Retainer* Agreement. These agreements are legal contracts that outline the terms of your agreement with the lawyer. Common terms in a retainer agreement include scope of representation, communication expectation, and fee structures. Scope of representation is important, it defines what matters this lawyer will (and will not be) helping you with. For example, if the scope of representation is drafting a will, the lawyer is responsible to draft your will, but not to act as your defense lawyer in a criminal proceeding against you. Lack of communication is one of the ways relationships fall apart, and the lawyer-client relationship is no different. Typically a retainer agreement will outline general expectations of communication; for example it might say "lawyer will endeavor to respond to client communications within two business days." Finally, the retainer agreement will outline the fee structure and any prepaid retainer fee requirement. A prepaid retainer fee is usually required by attorneys before they start work on your matter. This prepaid retainer fee is typically put into a separate bank account, sometimes called a client trust account, and the lawyer transfers funds out of the trust account into the law firm account only when they have earned the fees (which means when they have performed the work). Lawyers who do not require clients to prepay often don't get paid, or have to spend a lot of time and money chasing clients for unpaid bills. This is why many lawyers require clients to pay up front. The nuances of retainers, such as replenishment of funds or retaining a lawyer for general business matters, are beyond the scope of this text.

Now that we've covered the basics of hiring and interviewing a lawyer, let's talk about how to keep costs down.

Keeping Legal Costs Down

There are a few tricks to keeping legal costs in check: educate yourself about the law, be prepared for your meetings by thinking about what you want, and bring the necessary information to get the job done.

Educate yourself about the basics. Many lawyers cringe at the thought of their clients doing internet research about their cases, but for preliminary, basic things it's actually much easier for both the client and the lawyer if the client isn't completely clueless.

You don't need to read and know the law itself, but if you're interested, here is a quick tutorial of how to find it. For example, if you think you want a lawyer to write a will for you, before you go to the first meeting you could read your state's law about wills and intestacy. In order to find the law, you could start with three search terms: the state or jurisdiction you want to learn about, for example *Illinois*; the name of the law or body that writes the law, for example *statute* or *legislation*; and the legal topic you're interested in learning about, for example *intestacy*. With a Google search of *Illinois statute intestacy* the first result was the Illinois statute that covers the topic. Look for a .gov or another official suffix; there will be many unofficial sites trying to attract you. Statutes or laws are often organized in sections, so if you find the intestacy section, you should be in or near the section about wills and trusts.

If you want to really learn about the law, try going to a law library. Every law school that accepts public funds in the USA is required to open its library to the public. In addition to law schools, in larger cities there are often municipal law libraries that are open to the public. Law libraries are fantastic because they have both resources and highly trained librarians who can help you locate the information most relevant to your legal issue. Law libraries often provide free or low cost access to expensive online legal databases such as Westlaw or LexisNexis.

If you prefer online resources accessible from home, consider your community legal aid organization. You can find them by using search terms like "legal aid" or "pro bono" and the name of your local community. For example, a search of "Illinois legal aid" delivered a link to Illinois Legal Aid Online [https://www.illinoislegalaid.org/] which provides forms and how-to guides for common legal issues. The published resources are free for the public, and while they're not guaranteed to be the most current information, they're generally high quality and well written. Alternatively, you can often find information designed for the public on bar association websites. Finally, Harvard [https://guides.library.harvard.edu/free] has an extensive compilation of free legal research resources.

The goal is not for you to be your own lawyer; not even lawyers should be their own lawyers. The goal is for you to understand the process and what your lawyer will be doing, so that you can help expedite the process. By understanding the basics, you can anticipate questions your lawyer might ask, and be prepared to answer them. This leads to the next thing you can do to reduce costs: be prepared for your meetings.

Be Prepared

When meeting with your lawyer about estate planning, you'll need to provide information about your assets, your family, who you want to get what, and under what terms (if any). If you haven't really thought about how you want your assets distributed before the meeting, you'll waste precious time (expensive billable time) talking about general issues rather than specific options. Your lawyer will probably give you an asset disclosure form; be sure you take the time to complete it thoughtfully. For example, you might want your daughter, who is currently in medical school, to inherit everything except for your financial assets, which she will inherit after she turns 35 or finishes medical school. These are your assets, your loved ones' assets. It's the right time to give their needs and your wishes full attention.

Thinking about these things will probably be a highly emotional experience. You might feel anger, fear, resentment, love, joy, peace, confusion; and you might feel them all at the same time. It is completely normal to feel all of these things and more. It's normal to feel conflicted between how you think you should feel and how you actually feel. Inheritance planning is one of the few times in our lives that we simultaneously consider our legacy, our entire life experience, and all of the significant relationships throughout our lives. If you feel anger, resentment, or fear, try to acknowledge it and figure out why. Frequently, people who express joy or satisfaction in cutting someone out of their will are actually very hurt. If, during this whole estate planning process, you can identify unresolved issues, you might have an opportunity to work through them and/or mend the relationship while you're still here. And depending on the situation, that could be the best gift you could give to yourself and your loved ones.

You may have heard the old adage "a lawyer who represents himself has a fool for a client." This is true for lawyers and for most laypeople. It's difficult to see things objectively, from the viewpoint of a disinterested third party, which is precisely who will be trying to interpret your estate plan. It's usually a bad idea to write your own legal documents because it's difficult to write legal documents well. The medical school example above is illustrative of why. Let's look at the sentence again: *your daughter, who is currently in medical school, to inherit everything except for your financial assets which she will inherit after she turns 35 or finishes medical school.* The intention may seem clear but is it? What does the word finishes mean? Does that mean successfully completes school, finishes residency, and actually becomes a practicing MD? Probably not. Does it mean she must complete the program, even if she is last in her class? You and your daughter may have a shared understanding of what it means for her to *finish* medical school, but the court, trustee or a disinterested third party will not. Which brings us to the next point: read your estate planning documents carefully, and ask questions until you actually understand how everything will work.

Your lawyer will not suffer nearly as much as your loved ones will if there are errors in your estate planning documents. Read them carefully and do not expect that your lawyer got it right the first time; it is your responsibility to make sure your documents are accurate and correctly reflect your wishes. My father loves to tell the story of when he hired a lawyer to do estate planning. At that time, family benefit trusts were all the rage. Knowing this he specifically told the lawyer, "I have a non-traditional family. Do not draft a family benefit trust for me, because it will not work for my situation." A few weeks and a couple of thousand dollars later, the lawyer proudly delivered the finished estate planning documents. I bet you can guess where this is going and you're right. The lawyer put together a 12-page family trust, ignoring all of my father's instructions. Of course my father didn't sign them and immediately fired the attorney, but he still has those documents as a testament to always carefully reviewing documents you sign, especially those as important as your estate documents.

Firing Your Attorney

Don't be afraid to fire your attorney. Your attorney wants to help you and believes they can, otherwise they would not have accepted you as a client. However that's only one side of the relationship. If you're not happy with the relationship you have a right, no, a duty to end it. The idea of firing someone might be scary, especially if you've never done it before and you like to avoid confrontation. Remember, even though estate planning is highly personal in content, the relationship you have with your attorney is usually all business. You've hired them to perform a service. If you're not sure how to end the relationship here is some language that might help: "Thank you for all of your hard work so far but I've decided to terminate our attorney-client relationship effective immediately. Please provide a full accounting, refund my unused retainer (if any), and provide a copy of all of the documents relating to the matter to:" (Add the mailing address or encrypted file storage service location). Know that once you end the relationship, you probably won't be able to retain this lawyer again, so don't make it an impulse decision.

Keeping It Fresh

Time is the enemy of planning. As time passes, things change: your assets, technology, relationships with your loved ones and helpers, and the quality of your storage locations. Think of your plan like a beautiful apple on your kitchen counter. After three or four days, it's still perfectly edible but may not be as crunchy and fresh as it was. In a week you'd only eat it if you were starving. In four weeks, eating it might kill you.

Your inheritance plan is most fresh, relevant, effective, and complete the day you create it. A month or two later your plan may not be as fresh; perhaps you bought and sold some assets and got a new phone. Six months later, perhaps a third of your inventory is out of date and you've moved assets to a hardware wallet (which you have properly backed up, right?). With the right helper and a bit more hard work, your plan will probably still convey most of your cryptoassets to your heirs—but it's beginning to show signs of aging. A year later, your inventory will probably be significantly out of date, your relationships with your helpers might have changed, and maybe your heirs are all completely into cryptoassets and all have wallets they're using on a daily basis. Okay, I'm joking about that last one, but we can dream, right? In all seriousness, after a year your plan will probably be in need of a major refresh.

You have to find the right balance between the effort required to refresh your plan and the urgency of the refresh. Most people will only update their plans once or twice a year. If you try to do it more often, you'll probably procrastinate and not update it at all.

Inheritance planning is a holistic process. While we created the plans a piece at a time, in different sections of the book, these pieces came together to create a complete plan. When you update the plan, you must update the entire plan. Do not simply update one piece without considering how that update will affect the rest of the access plan, and the legal plan if you have one.

In order to ensure you update your plans regularly, take time now to set a calendar reminder for six months and one year from now. At the six month interval, you can make the decision as to whether or not it's the right time for you to update your plans.

Conclusion

Bitcoin, open blockchains, and asset tokens will change the way we do everything, including law and inheritance. These borderless technologies provide a reason to rethink legal systems and create new technology-based processes that can incorporate the ideals of fairness, justice, and empowerment to people. That is why I've been working exclusively with bitcoin, open blockchains, and smart contracts since 2014.

I am fascinated and excited by the intersection of law and technology. I continue to use, study, and experiment with cryptocurrencies, tokens, and smart contracts on a daily basis, looking for problems to solve and solutions to build. One of the issues I discovered was the lack of high-quality, pragmatic information about inheritance planning for cryptocurrencies and asset tokens. That is why I've written this book. As someone with practical and legal experience in the field of inheritance planning, I am excited by the promise of a future incorporating technology into these processes, but I'm also extremely skeptical when I hear people say that all the of problems have been solved with technology. You should be too.

Today, most cryptoasset owners have not made a plan for their loved ones to inherit their assets, and that's a real shame. If you are one of them, Get It Done! Every single person who owns a cryptoasset is a risk-taker; your loved ones should benefit from the risks you've taken. Our community needed a simple, practical guide to cryptoasset inheritance planning. I hope this book has given you the tools you need to make sure your heirs benefit from your foresight and diligence.

Want More?

Please Review This Book

If you enjoyed this book, please consider leaving a review on Amazon or wherever you purchased it. Reviews help people find the book and understand whether or not it's the right book for them. If you didn't enjoy the book or have suggestions, please contact us at info@merklebloom.com [mailto:info@merklebloom.com]. We hope you've found this book valuable and useful; if you haven't we want to understand why. We appreciate all feedback; your honesty is how we improve the book!

Sign Up and Keep Up

If you enjoyed this book and would like to receive news about the latest cryptoasset inheritance resources, learn about the next book in the series, and be entered into raffles for free or discounted copies of books in the series, please sign up to our mailing list.

We will not sell or share the list with anyone and we will not spam you. As a thank you for signing up, you'll be able to download PDF versions of all of the forms and worksheets in this book.

Sign up using the details on the next page...

To sign up please scan this:

Or type in this URL:

https://empoweredlaw.com/want-more-cip-p/

Or this shorter one:

http://bit.ly/wantmorecipp

Appendix A: Smart Contracts For Inheritance Planning

Can a smart contract handle the inheritance of my cryptoassets? What about a "dead man's switch"? Can inheritance problems be solved with a simple smart contract? Whenever I talk about estate planning and cryptocurrency these questions always come up.

The holy grail for cryptoasset inheritance is that an owner of cryptocurrency can use software to automatically transfer their cryptoassets to their heirs upon their death, without intervention by a court or any other third party. A will by smart contract. Sounds cool, right? Of course it does! No one wants to go to probate court. Even if you don't know what probate is, you've probably heard that it is something to avoid. Wouldn't it be easier, better, if we could just bypass the courts and all the potential disputes about who gets what with some sleek new smart contract software? Yes it would, and if that were possible today I'd be the first to advocate using it. But the truth is, this sort of solution isn't a solution at all. It's a problem generator.

Distribution of Assets Isn't the Real Problem

I don't say this because I'm a lawyer and want to make money from keeping your cryptoassets tied up in court. Right now, there are people who are investing their time and money trying to build these sorts of projects. Many of the projects are built by well-intentioned software engineers who have never actually dealt with inheritance relating to the loss of a loved one. Some of these projects have raised millions through ICOs (Initial Coin Offerings: a cryptocurrency equivalent of the more familiar IPO, and currently the most popular way to raise money for cryptocurrency and cryptoasset projects). Their websites are well designed, with links to academic-style lengthy white papers containing lots of fancy diagrams and complex descriptions of how cryptography works. For non-technical people, these whitepapers create the illusion that the founding team has solved all estate planning problems with software (their proprietary software) and that disputes about estates are a thing of the past. They view the problem as simply that of designating where assets should go. In fact, most of these products are nothing more than a software-based will. But those of us who have direct experience with these issues know that the real trouble doesn't come from process, it comes from people. It comes from the combination of grief, greed, and entitlement. And none of those problems can be fixed with software. Sometimes, software simply adds more problems.

Problems with Software-only Inheritance Planning

Let's look at the reality of software-only inheritance planning in a bit more detail. For now, let's forget that most of these platforms require your heirs to understand cryptoassets and how to use the "smart will" company's proprietary software. And let's ignore the glaring problems of centralized control over the assets, control over the smart contract asset distribution process, centralization of trust to the development team and all of their development choices. Or the issues of insecure data collection, insecure storage of identity, ownership, and assets. Or the likelihood that these companies will be targeted by thieves and provide a nice, software-based, opportunity for asset seizures. And let's ignore local laws, like those in Ontario, Canada that prohibit using electronic signatures for wills (special thanks to attorney Ana Badour for pointing that out). Even overlooking all of that, there are still glaring problems.

Oracle v Dead Man's Switch

The first problem most people recognize is: How will the

software know that someone is dead? Obviously you're not there to say, "Yes, I'm dead," so how will the software know to automatically transfer your assets? How will the software be able to verify proof of death? There are two ways most people try to solve this issue. The first is an oracle, and the second is a so-called "dead man's switch."

In this context, an oracle is an independent, third-party software agent that aggregates data (facts) and makes them available through a software interface. One example is an oracle that automatically accesses, and delivers upon request, all public record data for a specific geographic area—births, deaths, marriages and the like. In order to know if someone is dead, the inheritance software could ask this oracle to verify the death certificate of a named person with a specific date of birth. On its face, this seems like a viable solution, except that this requires you to trust the oracle, completely. If an error is made, your assets will be irrevocably transferred. Therefore, you must have complete trust in the data collection and delivery mechanism. What if someone else with the same name and date of birth in your area dies? Or, as once happened in a Mark Twain novel, the report of your death was greatly exaggerated. Or if there is an error on the death certificate that causes it to not match your information? What if the oracle's signing keys are compromised, perhaps by one of your highly-motivated heirs? What a juicy hacking target! Using this method also requires you trust that this specific oracle will not be unavailable (go out of business), thus unintentionally locking your funds into a smart contract that will never automatically trigger.

At some point in the future, maybe even in the next five to ten years, I believe oracles like these will become an important tool for automating *some* portions of estate distribution, but for now we cannot count on oracles to provide proof of death. This leaves us with the second option, the dead man's switch.

Instead of trusting an oracle to confirm that you're dead, you could instead confirm that you're dead by omission, by not taking some specific action. Your failure to act causes some other action to occur; this is called a dead man's switch. It's a fairly simple concept: every two weeks (or some other time interval set by you) you must check in to postpone activation. This usually involves logging into a website or responding to an automatic email request. If you check in on or before the designated time, then the clock resets and in another two weeks you'll repeat the same process. If you miss the check-in, or fail to complete the process, something happens. In the case of automated asset distribution, your assets will be distributed in the way you decided when you set up the service. The premise is that you will always check in, unless you're dead. Seems simple enough, what's the problem?

Let's test the premise. What are other reasons, aside from death, that you might not check in? What about natural disaster? At the time this book is being written, over 400,000 people on the island of Puerto Rico are still without power, months after a devastating storm, Hurricane Maria, hit the island. What about injury? If you are hospitalized for a serious injury or you have major surgery, you might not be able to check in on time. What about lost credentials? If you lose access to your email, phone, or laptop, how long will it take you to recover? Longer than your check-in time frame? What about bad actors? I know we don't want to think about this, but an automatic transfer of assets provides incentive for beneficiaries to keep you offline, to make it impossible for you to check in. Right now you're probably thinking, "Not my loved ones." And I hope you're right. But people change over time, and sometimes untreated addiction or bad choices in partners make people behave in unanticipated ways. The point is that there are many reasons aside from death that you might not check in. An automatic transfer of all of your assets based upon a simple failure to check in to an online platform is dangerous at best. I think it's better described as grossly negligent design.

Today the Tech Is Too Immature

But let's say we solve the "how to prove someone is dead"

problem. The technology is still not mature enough for long-term, absentee owner asset management. Remember, bitcoin was born in 2009, and ethereum in 2014. These technologies are still in their infancy and changing quickly.

In 2016, we saw "TheDAO" smart contract drained of millions of dollars because of a coding error that allowed someone to take more than their share of ether out of the contract. At its peak, this contract code controlled more than $150 million of value, and the social cost of this error resulted in more than lost money. The controversy over how to resolve the theft (and even whether it counted as theft) caused the ethereum blockchain to split; it's why today there are two: ethereum and ethereum classic.

In late 2017, more than $100 million was locked in a widely-used ethereum multisignature smart contract when the underlying libraries were accidentally "killed". As of March 2018, those funds remain locked. Isn't it ironic that your estate planning smart contract might die before you do? Oops! So when someone says they have a smart contract solution for estate planning today, I can't help but think they don't understand estate planning, smart contracts, or both.

Of course, at some point these kinds of solutions will be viable to handle many of the mundane aspects of cryptoasset inheritance distribution, and I'm looking forward that time. But to get there the software will need to be tested for years, with real money, in real inheritance situations. Money will be lost, lessons will be learned, and eventually the technology will mature enough to be able to deliver on the promises of smart contract inheritance management. But it's not ready yet and it won't be for many years. In the meantime, I know I don't want to be the test dummy. My guess is, neither do you.

When a Dead Man's Switch Might Be Useful

Even though a dead man's switch isn't recommended for

transferring your cryptoassets, it might be useful to communicate information that's not confidential. Google has made the dead man's switch a feature of Google accounts through the Inactive Account Manager feature. Once you set it up, Google gives you the option of deleting your account, allowing trusted people to download your data from some or all of the Google services you use, or simply sending a message to a trusted person. However, one of the challenges with this service is that it requires a phone number and SMS for identity verification. That's not ideal because you don't actually own "your" phone number; the phone company owns the number and porting someone's phone number is a known attack vector. If you want to learn more about this, feel free to read my free article about using two-factor authentication [https://empoweredlaw.com/articles/articles-2/].

This brings up an important concept, albeit one that is beyond the scope of this book. That is, what will happen to the rest of your digital assets when you die? There have been many great articles written on the topic and unfortunately the answers are far from certain. If you decide to hire an estate planning lawyer, they should be including your digital assets in your estate plan. If you've decided not to hire a lawyer, you should still consider if and how you want these assets to be passed on.

There are other simple dead man's switch email services, unconnected to Google, that promise to deliver a simple email message to recipients if you don't check in. All of them warn against using them to deliver confidential information. While most of my clients choose not to incorporate a dead man's switch into their access plans, there are some people who choose to use them. For example, they'll use a dead man's switch to tell their loved ones that they *have* cryptoassets and have left instructions for them at a certain location. But they never put confidential information like keys, seeds, or access codes in dead man's switch emails because unencrypted email is not secure enough for storage and delivery of sensitive information.

Many of today's password managers have some sort of inactive account authorization mechanism. Dashlane, Keypass, and Lastpass are some of the most popular with this feature. Assuming your heirs could legally access your password manager, you could leave a note for your loved ones there. You'd need to be sure they could easily find it when they finally get access and consider leaving a note in your will or access plan documents to let them know access is coming. If you're not yet using a password manager, you should be. If you want to learn more about using a password manager, feel free to read my free article about how-to use password managers [https://empoweredlaw.com/articles/articles-2/].

Smart Contracts in Inheritance Planning Today

Before you accuse me of being a Luddite lawyer, afraid of technology and of the future, ignorant of the incredible potential of smart contracts, allow me to clarify. Just because smart contracts aren't very helpful for cryptoasset inheritance planning today doesn't mean they'll never be. Quite the opposite; the ability to define complex spending criteria combined with the absolute guarantee of an immutable blockchain will radically reshape the way we control our own assets, and thereby radically reshape the way we plan for inheritance and succession.

The key issue is reconciling the time it will take for these technologies to reach maturity with the planning horizon of decades that is required for successful inheritance planning. Your inheritance plan may lie dormant for a very long time before it's activated. For smart contracts to become mature enough to support such applications they have to undergo rapid evolution. We can already see this in the frenetic pace of change in the Ethereum blockchain, where the dominant culture is to innovate, move fast, and if necessary break things.

There will come a time when the pace of change will slow, when the foundational smart contracts will seldom change, and will be well-tested and mature. Then, and only then, will we be ready to use these advanced technologies for solutions as nuanced, complex, emotionally laden, and fraught with legal pitfalls as inheritance and succession planning.

So what *can* we do with smart contracts, e.g. ethereum, scriptable spending conditions, e.g. bitcoin, and robust decentralized oracles? It turns out, quite a lot.

In both bitcoin and ethereum, it is already possible to create complex spending conditions which combine boolean logic, time-based controls, multi-signatures, and other blockchain primitives. For example, in *Mastering Bitcoin*, Andreas M. Antonopoulos illustrates an advanced bitcoin script which gradually modifies the spending conditions as certain deadlines expire. In the example there are three different spending conditions that activate at different times. In the first 30 days, the funds can only be spent with a 2-of-3 multisignature scheme. After 30 days, the funds can also be spent by a multisignature 1-of-3 scheme plus an additional signature by a third party (an attorney). Finally, after 90 days have elapsed, the attorney can spend the funds with a single signature.

Essentially this is a complex dead man's switch, where the participants in the multisignature scheme can reset the clock at any time by executing a transaction that rolls the funds into a new account. If they fail to act, additional spending options become available with a broader set of participants. This allows them to resolve problems such as loss of keys or incapacitation. While this example demonstrates the power of advanced scripting and could be helpful for short-term funds management, even in bitcoin where things move slowly this may not survive the time frames of inheritance planning. Keep in mind that this kind of advanced scripting has been possible since 2015 or 2016, yet not a single wallet or application has really exploited these capabilities and made them available to end users. It may take years before you see user interfaces or wallets based on compound conditionals, timelocks and advanced multi-signature schemes. Then the testing begins...

Ethereum allows even more complex spending conditions, limited only by the imagination of the programmer. Whereas there has never been a loss of funds in bitcoin caused by bugs or protocol changes (only by loss or compromise of keys), ethereum has had several such mishaps in its even briefer history. With great power and flexibility comes great risk and responsibility. Time and real-world testing, with real-value, on public networks will eventually allow us to trust these projects with more and more value. Eventually, perhaps, even with smart wills!

Appendix B: Acknowledgements

Thank you to all of the people who have trusted me to help them with their planning over the years. Thanks, too, to the many individuals who reached out and shared their stories of loss; these stories have helped to shape the book and its contents.

Writing a book is hard, not simply from a word count point of view, but also because it's a vulnerable thing to do. It's scary to put your words, your work, your legacy out there — even when your work is high-quality, even when you're considered by others to be an expert in the field.

As a first time author, it's a struggle to actually publish because you have to accept that your book isn't perfect. No book is. But having friends and colleagues review and critique the work makes it better. And making it better makes it less scary. Thanks to the following people, who have taken the time to provide feedback, critique, and generally have made this book better with their contributions (listed in alphabetical order):

- Andreas M. Antonopoulos
- Ana Badour
- Jessica Levesque
- Jameson Lopp
- Rodney MacInnes
- Brooke Mallers, PhD
- Amber D. Scott
- Dan Serlin
- Reuben Thomas
- Jeff Vandrew Jr.
- Walter the Unsalter

Of course, while those people helped to make this book better, there are people who have made this book possible, through their love and support. This book is dedicated to them because without them there would be no book.

To my father, who continues to teach me about people, relationships, family, and myself. We talk for hours, yet it feels like minutes. He *is* an entrepreneur in every sense of the word and has been since birth. I *am* my father's daughter and proud of it.

To my mother, who continues to teach me about compassion, empathy, and that it's never too late to make a change. She, too, is an entrepreneur; growing up she made me believe that I could do anything. I'm a better person because of her and I cherish our relationship.

To JD, who has kept her heart and a room open for me throughout my life. From you I learned that we can choose our family too.

To JL, who has taught me what real friendship looks and feels like. I'm constantly inspired by you. My life is better with you in it.

To my partner, the love of my life, my best friend. You believe in me, even when I don't; you've changed my life and make the impossible seem possible. You are a gift; I'm grateful for every single minute we share.

Finally, this book is dedicated to *you* — the person reading this book. If you're reading this section, you're both persistent and meticulous; you're the reason I've written this book. Thank you.

Appendix C: Keys, Seeds, and Access Code Templates

ASSET ACCESS CODE: do not discard

Software: ______
Date: ________________ Additional Code Req: Y / N
Assets: ______________________________
Notes: ______________________________

1		13	
2		14	
3		15	
4		16	
5		17	
6		18	
7		19	
8		20	
9		21	
10		22	
11		23	
12		24	

Figure 4. Template Seed

ASSET ACCESS CODE: do not discard

Software: ____

Date: ____________________ Additional Code Req: Y / N

Assets: ____

Notes: ____

Password/Passphrase: _________________________________

____

____

____

Figure 5. Template Access Code

Appendix D: SURE Evaluation Template

Table 11. SURE Evaluation Template

	Security	Usability	Resiliency	Efficiency
My Goals				
Quick Analysis				
Keys & Codes				
Places				
People				
Current Plan				

Appendix E: Letter to Loved One Template

Date:

Dear loved ones,

I'm writing this letter to let you know that I have cryptoassets that may be worth something. Please read this letter carefully and completely before taking any action. These assets are different from other assets—once these assets are transferred, there's no way to get them back.

Below is a list of people I trust to answer questions and help you through the process of finding and transferring these assets. Contact the people listed; do not trust only one person to handle this process. Watch all helpers very carefully, even the people listed here. Anyone can make mistakes, so make sure you understand what they are doing as best you can, and don't be afraid to ask questions and verify answers yourselves.

People who can help answer questions and guide you through this process are:

Insert your helper's names, organizational affiliation (if appropriate), contact information, and how they should verify identity (e.g. Keybase, photos)

Below is a list of devices, software, and assets I use to access these assets. Please put all of the devices away, under lock and key; store them securely until the assets have successfully been transferred to my heirs. Do not let anyone access them without supervision.

Insert your cryptoasset inventory here

Other notes:

In closing:

Checklist

Did you include information for:

- ☐ Helpers: names, contact information, photos if possible
- ☐ Devices: your phone, computer, hardware wallets, paper wallets
- ☐ Wallets: all of the software you use to access your assets
- ☐ Assets: include a list of assets
- ☐ Exchanges: be sure you've listed all of the exchanges that are holding funds for you
- ☐ Access: the information they'll need to find your storage locations and all necessary access codes

Made in the USA
Middletown, DE
27 June 2018